SECOND EDITION, REVISED AND EXPANDED

WHAT EVERY ENGINEER SHOULD KNOW ABOUT

MICROCOMPUTERS

Hardware/Software Design
A Step-by-Step Example

WILLIAM S. BENNETT
CAE-Link Corporation
Binghamton, New York

CARL F. EVERT, JR.
Department of Electrical
and Computer Engineering
University of Cincinnati
Cincinnati, Ohio

LESLIE C. LANDER
The Thomas J. Watson School of
Engineering, Applied Science, and Technology
State University of New York at Binghamton
Binghamton, New York

MARCEL DEKKER, INC.   New York • Basel • Hong Kong

Library of Congress Cataloging-in-Publication Data

Bennett, William S.
   What every engineer should know about microcomputers :
hardware/software design, a step-by-step example / William S.
Bennett, Carl F. Evert, Jr., Leslie C. Lander. —2nd ed., rev. and
expanded.
      p.   cm. —(What every engineer should know ; v.27)
   Includes bibliographical references and index.
   ISBN 0-8247-8193-7 (acid-free paper)
   1. Microcomputers. I. Evert, Carl F. II. Lander,
Leslie C. III. Title. IV. Series.
QA76.5.B398   1990
004.16—dc20   90-49152
                                                    CIP

This book is printed on acid-free paper.

Marcel Dekker, Inc.
270 Madison Avenue, New York, New York 10016

Current printing (last digit):
10 9 8 7 6 5 4 3 2 1

PRINTED IN THE UNITED STATES OF AMERICA

PREFACE TO THE SECOND EDITION

In this second edition, chapters have been added to addresss what is becoming an increasingly important aspect of microcomputer work. Often, changes must be made to the application, perhaps years after the original work was done, and by entirely different people. If the original application is done in the right way, the places where the changes must be made can be more easily found, without intense research into the original problem, and the changes more easily designed and inserted. This has been the focus of the U.S. Department of Defense's software initiatives over the last decade or more, and we have elected to illustrate these principles by applying the DoD software language "Ada" to the example used in this book to illustrate the goals and design methods that should be adopted by the designers, whatever software language they choose. The Ada example will, we hope, guide designers in selecting a language that has as many of these change-related features as they believe the complexity of their work demands.

William S. Bennett
Carl F. Evert, Jr.
Leslie C. Lander

PREFACE TO THE FIRST EDITION

This book is intended to serve the needs of a large class of engineers, managers, sales representatives, technicians, and students who want to gain an insight into the complexities of microcomputer applications, but who do not work with these devices every day. No special knowledge about microcomputers, or even computers, is assumed; technical information is supplied as it is needed throughout the text.

The presentation is built around a simple example, the measurement of the number of gallons of liquid in an oddly-shaped tank, employing a float and a microcomputer. As the solution is developed, facts about microcomputers are brought in, each illustrated by a diagram placed directly next to the idea it illustrates.

The details of the solution are given down to the level of the individual assembly-language instructions and the individual electrical connections, to give the readers a good appreciation for the complexity involved in a microcomputer application.

No attempt is made, however, to treat all of the instruction formats of the selected microprocessor, nor all of the techniques of their use; nor are all of the available auxiliary electronic devices explored. In a similar way, the book does not attempt to cover a range of different microprocessors. The aim is to show newcomers to the field how a

particular problem would be solved, and to impart to them the knowledge they would need in order to solve the problem, thus giving him or her, in a relatively pleasant and painless way, (1) a grounding in a representative set of techniques, (2) a clear idea of what a microcomputer designer or application engineer must go through, (3) an understanding of what must be communicated to a microcomputer development team, should they be in a position to use such services, and (4) enough real understanding of the process so that, if they must obtain for themselves a much larger body of knowledge about microcomputers, they will be better prepared to find, select, and assimilate it.

William S. Bennett
Carl F. Evert, Jr.

CONTENTS

WHAT EVERY ENGINEER
SHOULD KNOW ABOUT
MICROCOMPUTERS

1 INTRODUCTION

In just a few years, the microprocessor industry has grown so much, and the cost of microprocessors has been so drastically reduced, that these devices are now within the reach of every engineer, technician and instrumentation salesman. Almost every new instrument, controller, or data collection system has a microprocessor in it. Engineers of all disciplines, and many other people in technical work, find that they have to have a good understanding of the basic principles of these devices. Fortunately, the field is open to anyone willing to learn a few simple principles, and it is not hard if you spend your time on the right ideas — something this book will help you do. The world of small computers is easy to enter, and once there, it is fun.

You should think first of the job
the microcomputer has to do.
We're going to do that in this
book; once the job has been es-
tablished, we'll show you the details of how the microcomputer
carries it out. So the design starts
by figuring out what the job is —
in formal engineering jargon, it
starts with an analysis of the system which establishes the requirements, then proceeds to the implementation which accomplishes
the result. The steps are easy and
logical.

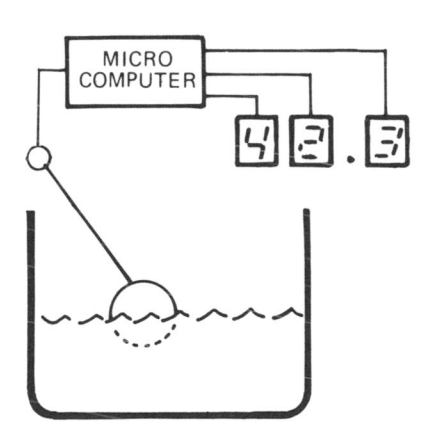

Suppose that the job is to report
how much of a certain liquid is
in a particular tank, and then display a number or a message that
tells how much there is. The tank
has an arm with a float on it,
which moves up and down as the
liquid level moves up and down.
What can a microprocessor contribute to the operation of this
simple system?

We're going to follow this particular problem in detail, all the
way to its solution, in this book.
But as we go through the steps of
the solution, we'll need to know
facts about microcomputers and
how they're used; so interspersed
with the solution we'll provide

some explanations of those facts, and we can begin with the way a microprocessor chip looks to the outside world:

The inputs and outputs of a microprocessor are voltages — either zero or five volts — that appear on several dozen terminals, or "pins," around its outside edge. So if we want to use any sensors as inputs, or displays as outputs, we have to match their voltages, and their arrangements of wires, to the microprocessor in some way.

And often there have to be some changes in the input information before it is passed on as an output. Inputs from sensors have to be converted to numbers or alphabetic messages for the output display. The microprocessor can provide an infinite variety of these conversions of input data to output data, but you have to specify exactly what they are. You do it with strings of simple commands, that are put into a permanent memory inside the microcomputer. These strings of simple commands, or instructions, are called programs, and programming is the major activity of thousands of engin-

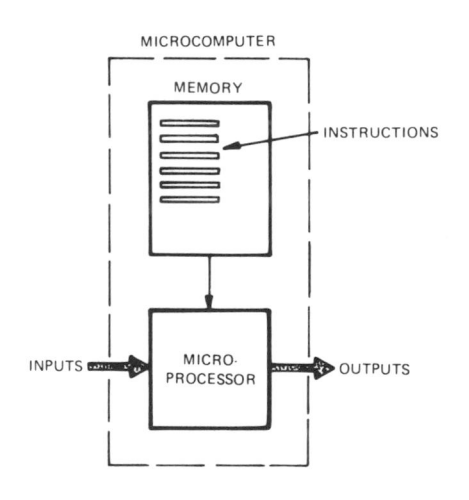

eers, technicians (and high school students) these days.

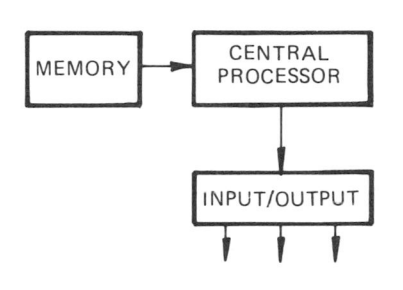

All computers, of course, have a central processor, some memory, and some way to bring data in and present it to the outside world. What is peculiar to the microcomputer is that the processor is usually on one chip, and perhaps memory and input/output circuits are on that same chip, but the memory and I/O circuits are at least in nearby small chips on the same supporting circuit board. Microcomputers are not only small but low in cost, at least relative to the larger computers. Typically, a microcomputer will be used where there is only one or a few different tasks to perform; where the string of instructions, or program, is mostly fixed; where the range of input/output equipment to which it is connected is not as wide as with larger computers; and where speed and efficiency are not as important. A microcomputer, for instance, would be found in a microwave oven, a laboratory instrument, a machine tool, or a cash register terminal: here there is no need for high-speed punched card readers, line printers, large magnetic tape or disk mem-

ories, and so forth. The input/
output is limited to a keyboard,
some digit displays, perhaps a
connection through a telephone
line to a larger computer somewhere. The program remains the
same because the task remains
the same: to cook food, to measure blood samples, to guide a
cutter, or to compute and print a
grocery tape.

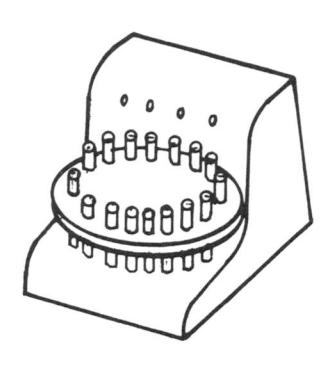

Larger computers, in contrast,
handle a generally wider range of
jobs. The "minicomputer" is the
next step up from a microcomputer; usually it is contained in a
box "the size of a breadbox,"
and has a printing console, perhaps a line printer, and one or
more magnetic tape units. It
would handle the paperwork for
a small office — payroll, inven-
tory, tax computations, sales
analyses, and other activities that
can change from day to day —
though it must be said that the
line between microcomputers
and minicomputers is blurring,
and some well-equipped microcomputers are encroaching on
some of these tasks.

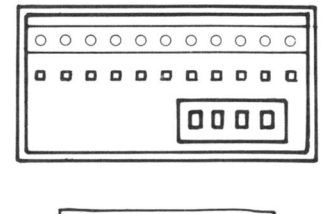

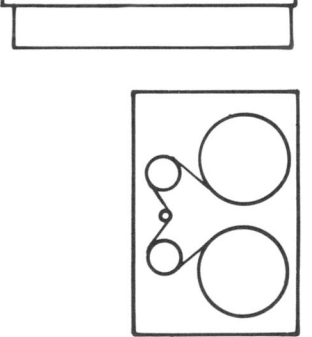

Larger still than minicomputers
are the "main frame" computers,
used, for instance, in large offices,

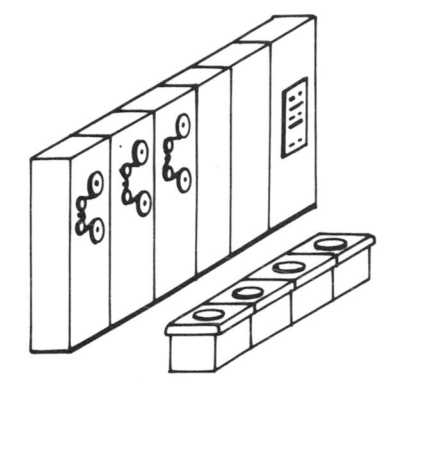

insurance companies, or the internal revenue service. These computers deal with very large files of data — lists of millions of insurance policies, or tax returns for an entire region of the country. Rows of magnetic disks and tapes will be connected to hold this data, and the input/output problem will be a major part of the task. Other large main frame computers will handle "scientific" computing — perhaps the enormous number of computations needed to produce a daily weather forecast for the nation. There, speed and efficiency in arithmetic, far beyond what is found in the microcomputer, are needed.

If you were to think now just of the microcomputer application — the task that is small, relatively fixed, not too demanding of input/output paths or fast, efficient arithmetic — what would you want in a device to solve the problem? What would you want as the important features of a box that would convert input signals to output information in your device or project or gadget? See if this description doesn't fit what you'd want:

It would be a very small box
with convenient mounting lugs,
very inexpensive and readily available in every small city. It would
have its own self-contained power supply. But most important,
it would have dozens or perhaps
hundreds of terminals on it, to
which ordinary wires are easily
connected, leading to any one of
dozens of kinds of sensors and
output devices: not only key-
boards, but temperature and pres-
sure sensors, liquid level sensors,
flowmeters, light detectors, and
so on; and not only automatic
typewriters or video screens but
electric motors, valves, solenoids,
lighted displays and actuators of
all kinds.

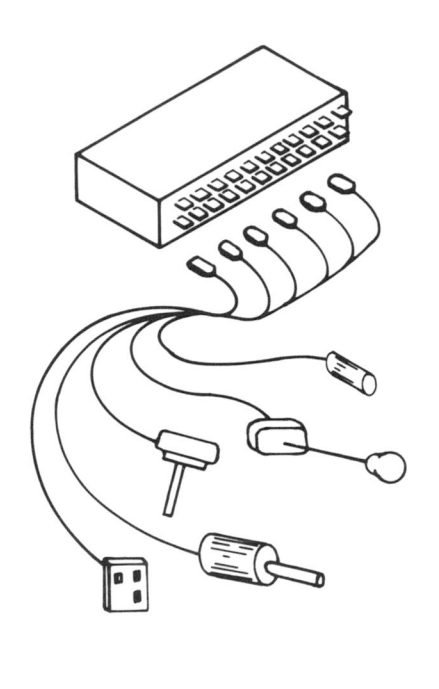

Some of you will be deep enough
into the computer world to want
the box to "talk" to larger computers, floppy disk units, magnetic tapes, remote terminals,
and other computer peripherals.

For the rest of us, those things
are a means to an end: just intermediate storage, for instance,
that we might need to get the
primary job done — storage we
might wish was right inside that
inexpensive little box.

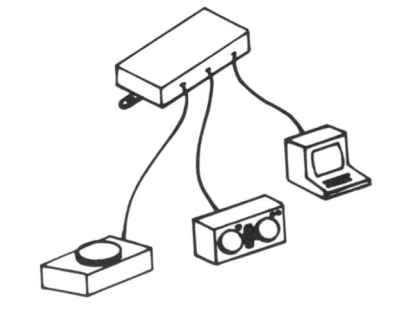

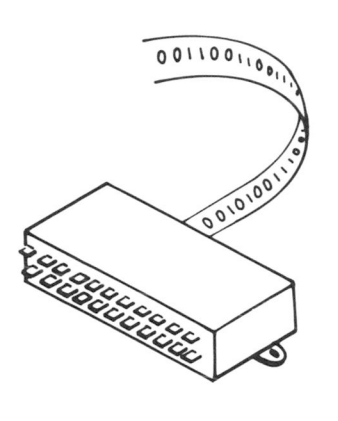

Most of us are aware that a micro-computer needs a long set of instructions to handle any useful task. Of course, what you would really want is a microcomputer that comes with some way to figure out those instructions, and get them into the box, with as little trouble — and as few mistakes — as possible; preferably in terms that make sense in your application, not some strange code that exists for the convenience of the microcomputer itself.

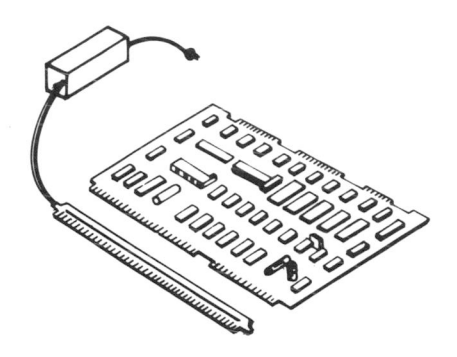

At present, though, these goals are not as close as we might hope. The usual microcomputer is still a fairly good-sized printed circuit board with dozens of components on it, needing a separate (and expensive) power supply — although, as we read continually, the heart of the system, a small "chip" called the microprocessor, is in fact packaged in a very small box. Half an inch by two inches by a quarter of an inch, it sells for anywhere from five to twenty dollars in the simpler models.

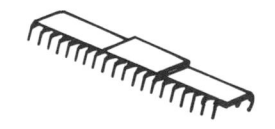

And, unfortunately, programming the microcomputer is still a very expensive, time-consuming, error-filled process, almost al-

ways carried out in a language far stranger (to any of our applications) than we would like.

Putting a microcomputer to work, for a number of years to come, will be an involved task demanding careful matching of electronic components, prudent packaging, and laborious programming and debugging.

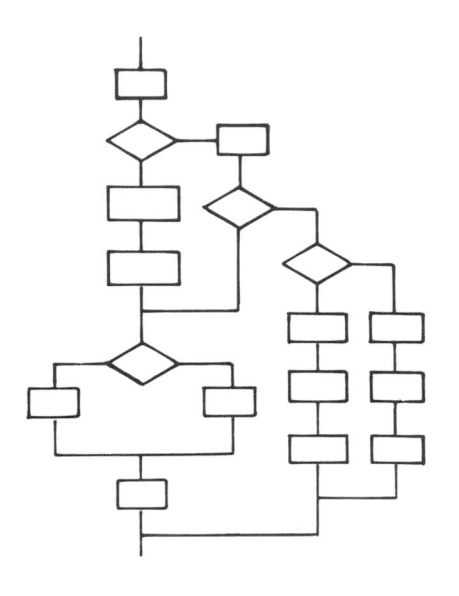

The major problem facing engineers and designers trying to employ microcomputers today — unless they are experienced in both electrical engineering and small-computer programming — is getting a good base of concepts so that the huge volume of manufacturer's literature on hardware components, and the hundreds of different publications on software development techniques, can become readable and usable. Most of this mass of written knowledge is so specialized that the casual reader is completely mystified; and most of the few books available on basic principles are either too simple to be useful, too complex to be readable, or require intense study in a classroom.

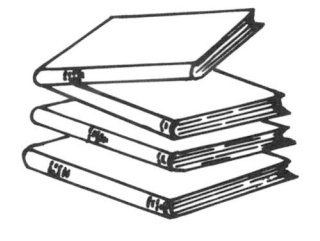

What we want to do in this short book is to provide you with this

BLOCK DIAGRAM

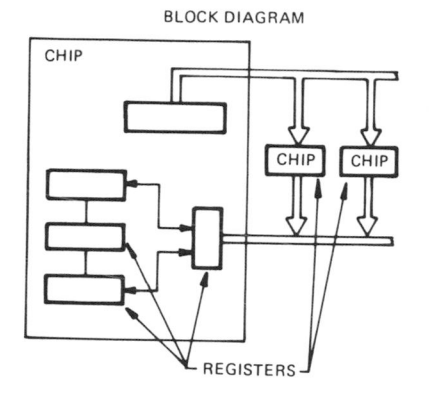

CHIP

CHIP CHIP

REGISTERS

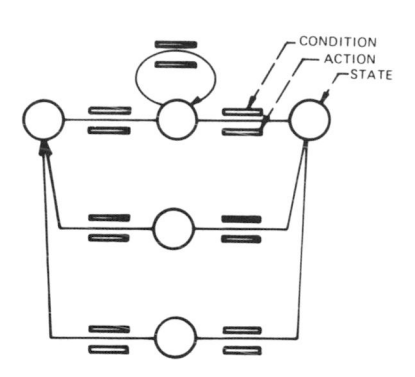

CONDITION
ACTION
STATE

MICROPROCESSOR

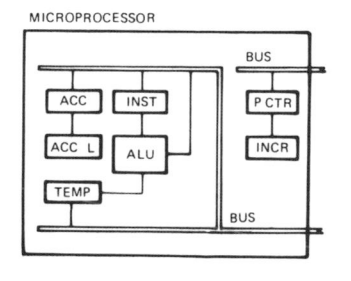

BUS

ACC INST P CTR

ACC L ALU INCR

TEMP

BUS

base of ideas, to make at least some of the literature understandable. And we want to leave you with a method with which you can work your way through a real application.

We're going to do this first of all through block diagrams that show the flow of information and electronic signals among the different components, both inside and outside the "chips," but particularly what we call registers: blocks that usually hold one number or some other small unit of information. We'll show how information is moved about from register to register, and processed or modified, with "state" diagrams, "Warnier-Orr" diagrams, and flow charts to make the program instructions clear. These "software" diagrams will be shown on all the levels you will need: from the top level matching your application, down through the operations at the input/output level; then the level at which the microprocessor "talks" to other components in the microcomputer; and what you will need to know on the level at which registers and arithmetic units interact inside the microprocessor.

In this small book none of this
can be done in anywhere near
the detail needed to make it un-
necessary for you to study any
other literature. But it will help
you organize your thinking and
form a framework in your mind
on which you can build, and into
which you can put the information you read in all those hundreds of other publications.

Finally, we are going to show you
what you will need to do if your
business requires changes to your
application, perhaps years after
the original work, that must be
done by perhaps entirely different people. Sometimes that means
that changes are not even possible
without doing the original work
all over again. But there are principles that can be used in the original work that will make the
changes not only possible, but
actually easy and cost effective.
These principles have come out
of the work of the U.S. Department of Defense, which had the
"Ada" programming language designed to embody them. Your application probably will not require Ada — it is a complex and
difficult language — but the principles can be applied with many
different languages and can in-

deed make a difference in the cost effectiveness of your design procedures.

2

A PROBLEM

Let's continue now with our simple example of a microcomputer application: one which is perhaps too simple to be economical at present microcomputer prices, but which illustrates some important principles — and might well be economical enough in coming years.

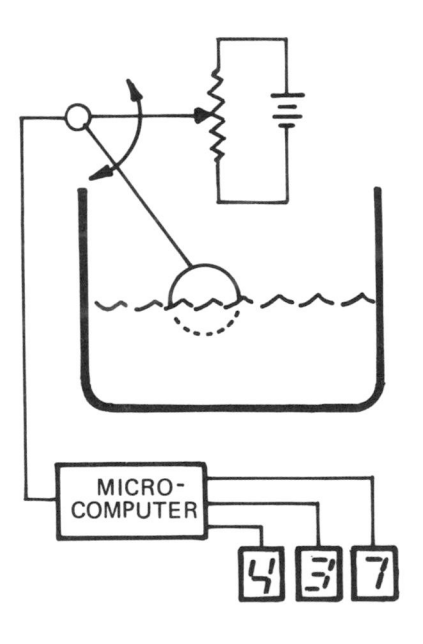

The job is to measure the amount of a liquid in a particular tank, which has a float on the end of an arm, which moves up and down as the liquid level does. The arm rotates the shaft of a potentiometer, or variable resistor, and so increases the voltage on the wiper of the "pot." The microcomputer's job is to convert this voltage into a reading on a set of digital readouts, of the type that have seven segments that can be lighted for each digit, and which will give the amount of liquid in, say, gal-

MICRO-COMPUTER

4 3 7

lons. Let's further say that the capacity of the tank is just under 100 gallons.

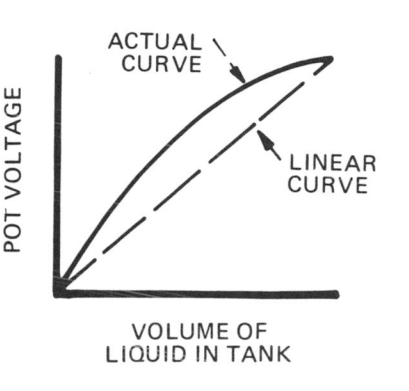

ACTUAL CURVE

LINEAR CURVE

POT VOLTAGE

VOLUME OF LIQUID IN TANK

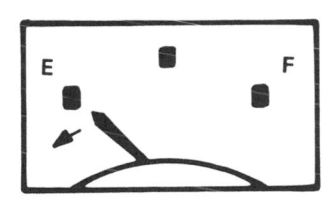

E   F

Applications of this kind often have complicating problems. In this case, the float and the "pot" do not produce a voltage that goes up linearly (that is, proportionally) with the number of gallons, or even the level of the liquid in inches from the bottom of the tank, because as the arm falls with the liquid, the shaft of the pot will rotate faster and faster; most gas gauges in automobiles, for instance, are built this way and fall much more rapidly as they approach "empty." So we want the microcomputer to compensate for this nonlinearity, and produce an output reading that accurately reflects the number of gallons in the tank.

The accuracy of measurement will necessarily be limited by the use of a potentiometer: although some are available with precisions up to 0.1 percent, when all effects are counted (including wear over time) one-half of one percent, or one part in two hundred, would be about what is realizable.

If the tank, then, contains 100 gallons, we would want a readout of three digits, to take us in 0.5-gallon steps from zero to 99.5 gallons.

3   A SOLUTION STRATEGY

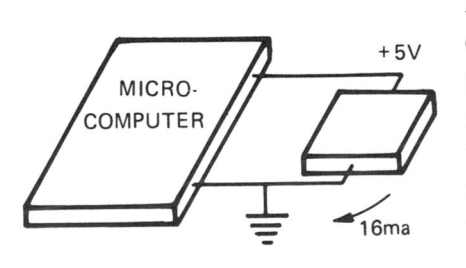

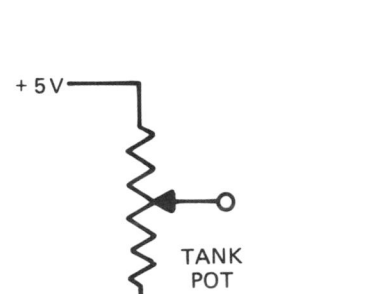

Because of its extremely low cost, we wish to use a microcomputer to perform the measurement. But we are measuring a voltage which varies smoothly (if not linearly) from zero to some voltage. Most microcomputers are capable of producing or accepting voltages of zero or five volts on their output or input lines, at a maximum current of about 16 milliamperes on each pin. To be generally compatible, we choose a voltage for the potentiometer that ranges between zero and five volts. Remember, though, that we said the microcomputer generates and accepts voltages that are EITHER zero or five volts; they are digital signals, and aren't supposed to take on values in between. (Digital signals, sometimes called logic signals or logic levels, do vary in voltage somewhat depending on

how much electrical load is connected to each circuit. A +5-
volt signal, for instance, may
drop to as low as 2 volts, yet be
considered to mean "ON," and
the 0-volt signal might rise to as
much as 0.8 volts yet be considered to mean "OFF." Often, for
this reason, the signals are called
simply "high" and "low." For
clarity, however, we'll continue
to call them "+5 volts" and
"0 volts").

So a way is needed to convert a
signal which varies smoothly —
like the potentiometer voltage —
into one which exists at only
those two states, yet carries the
same information.

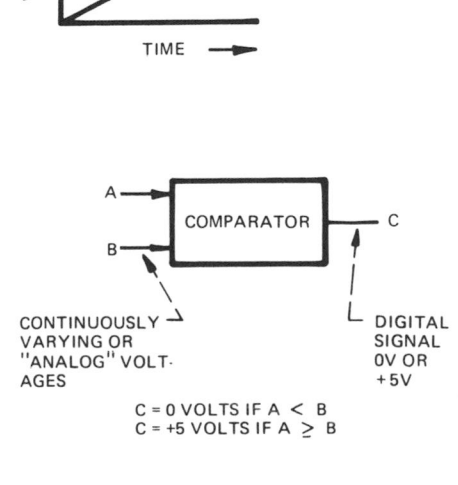

C = 0 VOLTS IF A < B
C = +5 VOLTS IF A ≥ B

A time-honored method of converting a varying level to an
on-off signal is by using a ramp,
or steadily rising voltage, com-
paring it with the unknown voltage, and turning on an output
when the ramp just surpasses the
unknown voltage. Part of the job,
then, can be done with a comparator circuit — one which com-
pares two voltages, A and B, and
generates an output which is off
— at zero volts — while A is less
than B, but comes on — at +5
volts — when A is equal to, or
greater than, B.

The ramp itself can be generated by an integrator circuit, which accepts an input signal which can be one set voltage (+5 volts, for instance) and produces an output that depends both on the input voltage and on how long it has been on. It generates, in other words, a linear ramp: a straight line on a graph of voltage versus time.

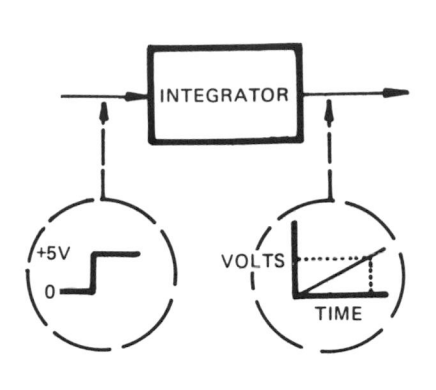

Of course, when the comparator signals that the gradually increasing ramp has just surpassed the pot's voltage, it's necessary to deduce what the ramp's voltage was at that time. The ramp rises very steadily with time, however, so by measuring the time since it started, a good number proportional to the voltage can be found. So one of the functions the microcomputer will have to perform is that of a clock.

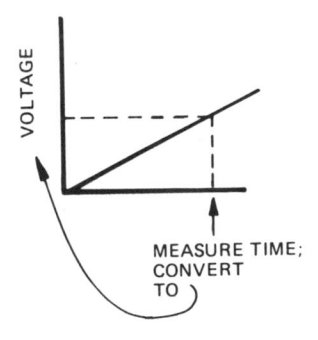

The clock function can be obtained by holding a number inside the microcomputer and increasing it by 1 at regularly timed intervals. Numbers can be held inside the microcomputer in what are known as "registers"— rows of circuits, each with one output, that can be ON or OFF (+5 or 0 volts). A register, then,

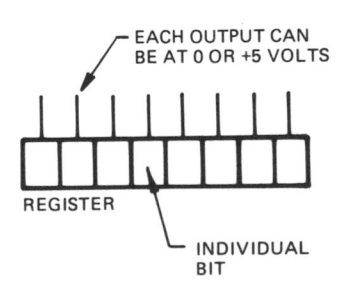

can hold any number that has first been converted into a row of ON/OFFs — that is, a row of bits. There is more than one way to represent a number as a row of bits, as we will see later in the book.

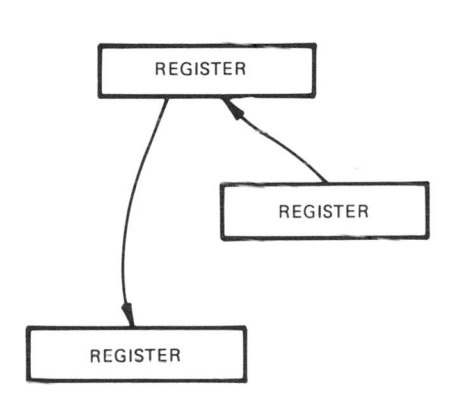

The idea of the register is very important in microcomputers. The great majority of operations inside the microcomputer are just transfers of data from one register to another, sometimes between registers of slightly different kinds, but still just transfers of data. Computer memory can also be thought of as large groups of registers. Today, now that magnetic core memories have largely been replaced by electronic chips, that is exactly what a memory is: a large group of registers.

Our strategy, then, will be to have the microcomputer, and these auxiliary circuits, generate a steadily rising voltage, keeping track of a time-or-voltage number in a register inside the microcomputer. Then, externally, that ramp voltage will be compared to that on the "pot," in the comparator, which reports back to the microcomputer when the

microcomputer's voltage just exceeds that of the pot's. At that moment, the microcomputer will convert the value in the register into an output reading, which it will send to the three displays.

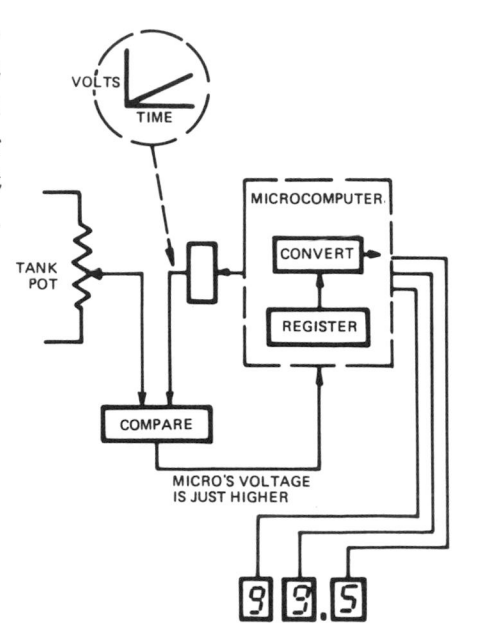

4

ENGINEERING
OUR SOLUTION

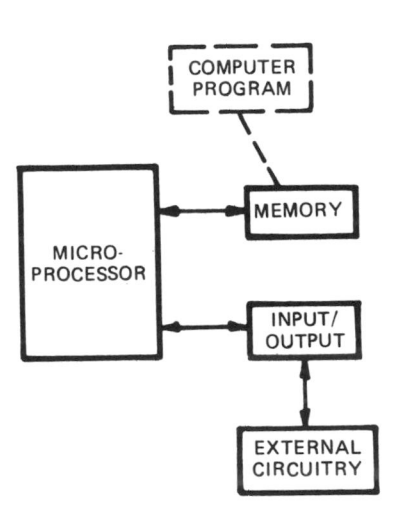

The first two steps in an engineering solution are to define the problem, and to plan a solution strategy. In Chapters 2 and 3, those are the steps we carried out. Past that point, we must develop the solution; that is, consider the ranges the variables must go through, adjust the speeds, voltages, and ways of representing numbers so that they fit these ranges, and devise economical methods of input, output, memory, and computation in our microcomputer. Electrical connections between chips and circuits will have to be worked out, and a careful breakdown of the processing steps made, in some systematic form of charting, so that the computer instructions can be developed in a straightforward way that is not too difficult to test. Let's begin with the external circuitry:

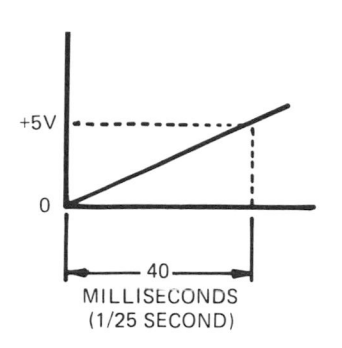

40
MILLISECONDS
(1/25 SECOND)

The rate at which the integrator's output voltage rises with time can be controlled with externally attached resistors, so that we can, for instance, arrange to have the integrator's output increase from zero to five volts in about 1/25 of a second (that is, 40 milliseconds). For almost all applications, that will be a very acceptably brief ramp to use.

At the instant that the comparator "fires" — gives the signal that the integrator output has surpassed the pot's voltage — we can know the ramp's voltage, as we said, by using the microcomputer to measure the time at which that event occurred. The microprocessor is usually sequenced by a crystal oscillator, of the same kind used in quartz watches; so the instructions are executed at a steady, known rate. If, using these instructions, the value in a register in the microcomputer is increased by 1 at regular intervals — say, every 200 microseconds or so — and the time remaining in each 200-microsecond period is "timed out" by an appropriate loop of instructions, then the count in the register will be quite as accurate a measure of time as any crystal-controlled clock.

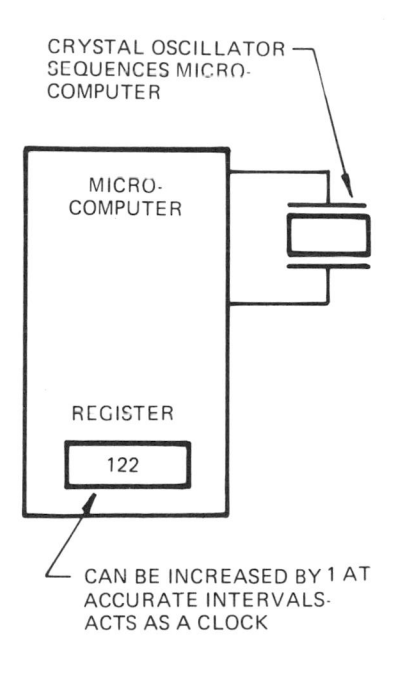

CRYSTAL OSCILLATOR
SEQUENCES MICRO-
COMPUTER

MICRO-
COMPUTER

REGISTER

122

CAN BE INCREASED BY 1 AT
ACCURATE INTERVALS-
ACTS AS A CLOCK

When the comparator "**fires**," the register can be consulted to obtain the number, which will be a measure both of the time and of the voltage. This is, in fact, the way that many analog-to-digital converters work.

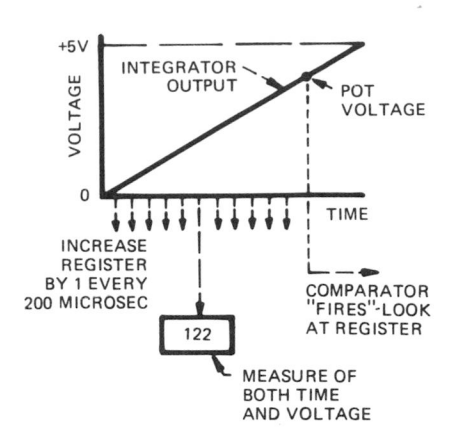

The microcomputer will "know" the voltage of the pot to a precision that depends on how many counts are put into the register for the full 5-volt range. If it does take the integrator 40 milliseconds — that is, 40,000 microseconds — to reach 5 volts, and if the register is counted up by 1 every 200 microseconds, then there will be 40,000/200 = 200 steps. That, as we said, would give us the number of gallons in a 100-gallon tank to a precision of half a gallon, or one-half of one percent.

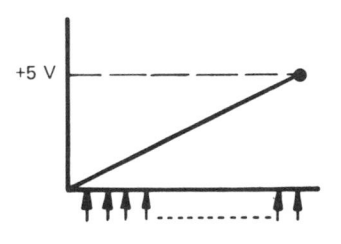

200 STEPS OF
200 MICROSECONDS
  = 40,000 MICROSECONDS
  = 40 MILLISECONDS
  = .04 SECOND

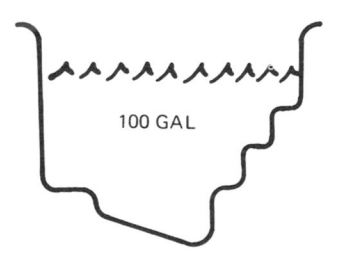

100 GAL

Now the number in the register must be converted to the number of gallons of liquid in the tank.

$$\frac{100 \text{ GALLONS}}{200 \text{ STEPS}} \quad \frac{1/2 \text{ GALLON}}{\text{STEP}} = 0.5\%$$

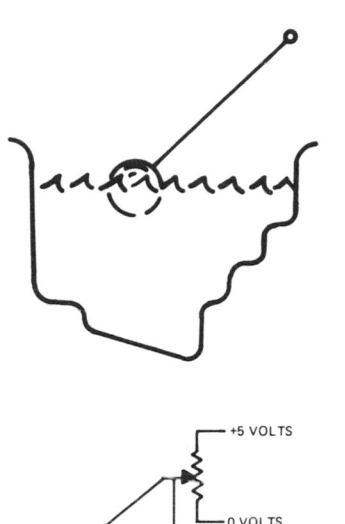

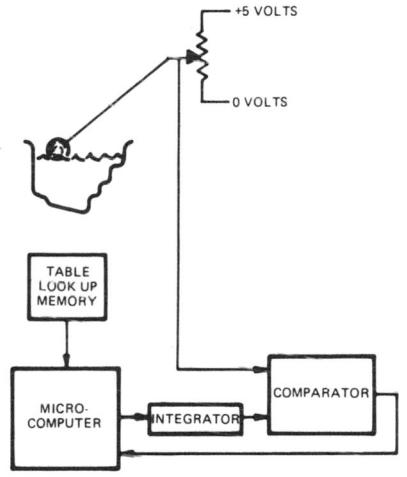

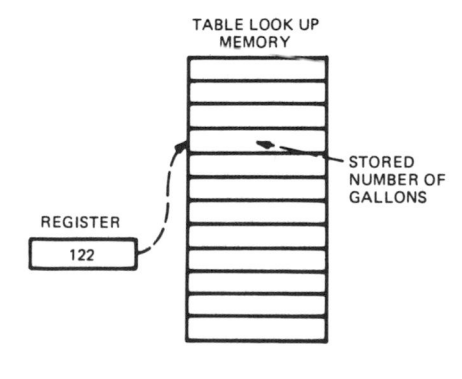

Let's suppose, however, that the tank is of some strange shape, which is probably the way most tanks are. That, in conjunction with the geometry of the float arm, will mean that the number of gallons of liquid in the tank will not be directly proportional to the pot's voltage, or to the number in the register. In fact, there will be some complex algebraic expression relating one to the other.

It is, of course, possible to derive the expression and show how it would be programmed. However, another method is often used in microcomputer applications, and we will use it here because it illustratrates a number of important ideas in the use of microcomputers. It is the table lookup method: a table of numbers of gallons, each corresponding to one of the register values, can be stored in a memory in the microcomputer. The contents of the register are used as the entry into the table, to find the correct number of gallons directly. The block diagram of the system can be expanded to show this memory.

The memory is actually just a large group of registers, in a special arrangement that allows us to choose one of them, and either transfer its contents into some external register (that is, "read" it), or transfer the contents of an external register into it (that is, "write" it). Except for the business of choosing the right register out of the memory, then, memory operations are really just more transfers from one register to another.

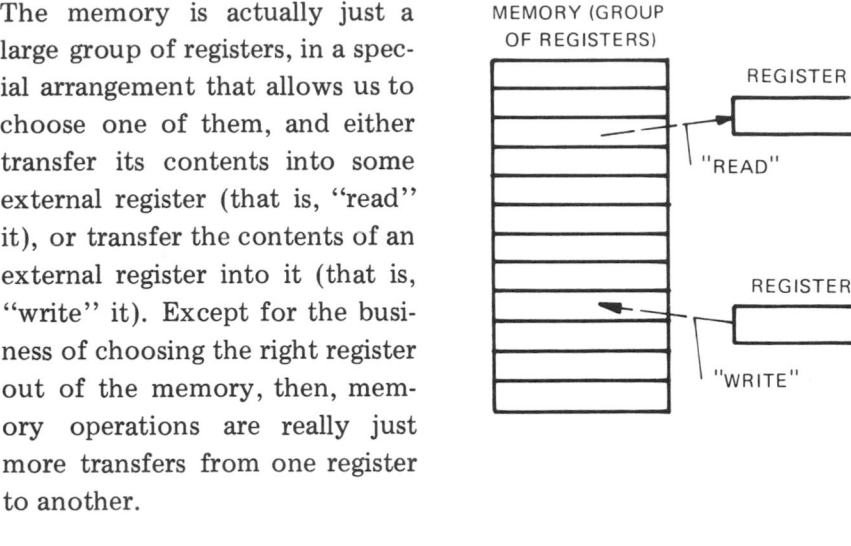

Before we're through, we will consider in detail how to hold that table in memory, how to use the register contents as the entry into that table, and how to get the answers out. We'll also consider what format the data will have to be in, in order to make the lookup as efficient and as inexpensive as possible, all things considered.

Now, however, let's go on to consider the other side of the problem: The digital readouts that have to be driven so that they read the number of gallons. Each one is composed of seven lighted segments, which have to be driven individually to be

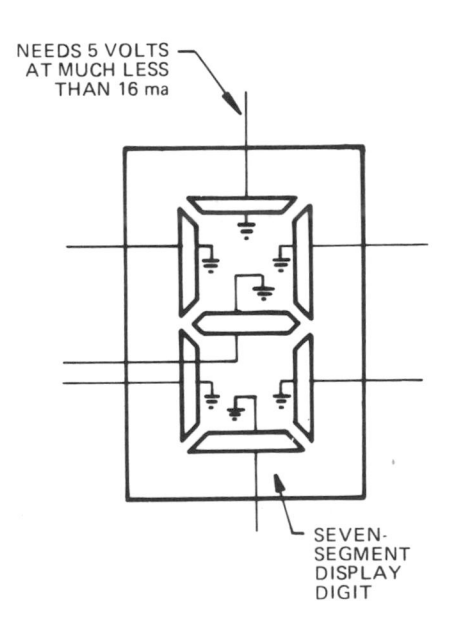

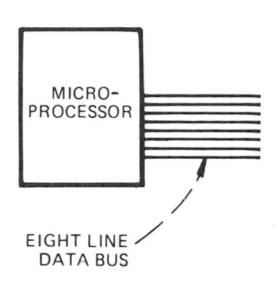

EIGHT LINE
DATA BUS

lighted, and so contribute to the display of the right numeral. For three digits, 3 x 7, or twenty-one lines have to be driven; displays are available now in which those lines can be driven with five volts, and which consume rather less than 16 milliamperes each. So on the surface it would appear that microprocessor outputs could be used directly. That would be very good, but most microprocessors do not provide twenty-one output lines; a more usual number is eight, and other microprocessors provide four or sixteen. So some means of passing the information out on (say) eight lines, and then catching that data externally, saving it and producing it continuously for the displays, has to be provided.

In microcomputers, that job is done by I/O (for "input/output") chips; they have eight lines in and eight lines out, and contain storage for eight "bits," that is, when presented with zero or five-volt signals on each of the eight input lines, and told to store the data, they can be thereafter made to ignore fluctuations on the input lines and present, on their output lines, the remembered set of data.

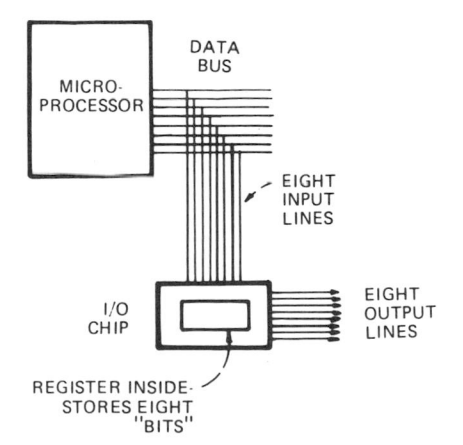

REGISTER INSIDE-
STORES EIGHT
"BITS"

Because the I/O chips do have storage in them, they can be considered as just more registers, which in fact they are — with some added functions. Ways exist to select them with addresses, so that they can be singled out to be sent data, or to be asked for data they are bringing in from the outside world.

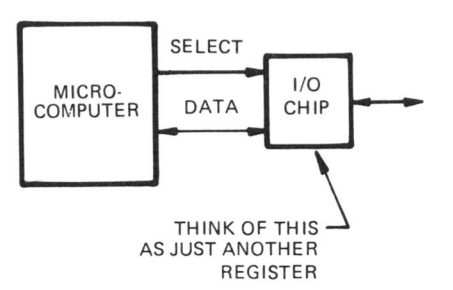

THINK OF THIS
AS JUST ANOTHER
REGISTER

We have twenty-one lines of data to send out to the three seven-segment displays. I/O circuits are usually made to handle eight lines; but it will be most inexpensive for us to use three of those circuits, one for each display, and just "waste" the extra line on each one. That's typical of many decisions that will be made in microelectronic equipment: the standard parts are so inexpensive that many otherwise "wasteful" practices are clearly the best choices.

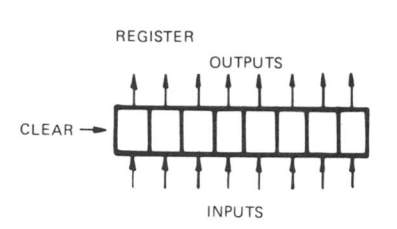

REGISTER

OUTPUTS

CLEAR →

INPUTS

The registers inside the I/O chips have input lines, as do almost all registers. There is one for each bit of the register. There is also a single "CLEAR" line for the entire register; when +5 volts is placed on the CLEAR line, all the outputs of the register come back down to 0 volts and stay there. They stay there, that is, until a signal (+5 volts) is placed on one of the input lines; then that bit's output goes up to +5 volts.

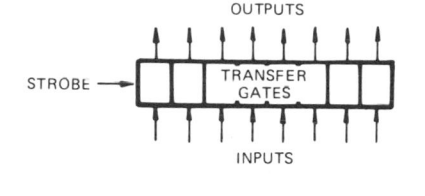

OUTPUTS

STROBE →   TRANSFER
            GATES

INPUTS

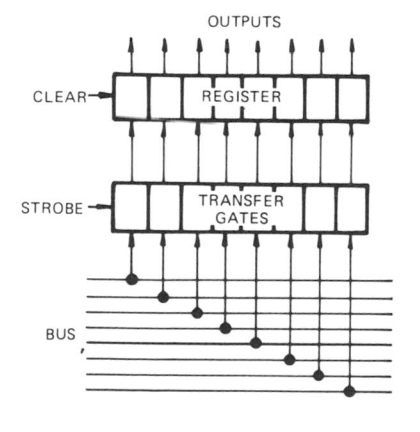

OUTPUTS

CLEAR →   REGISTER

STROBE →   TRANSFER
            GATES

BUS

Input lines to a register can be controlled by "transfer gates," another row of circuits. Data, possibly from a group of master wires (a "bus"), enters the transfer gates on a group of pins. When a control line on the transfer gate is momentarily pulsed ("strobed"), any +5 volt signals on the input are passed through to the output — and on to the inputs of the register. If the register had been cleared just previously, it will now contain that row of data.

I/O chips generally contain not only a register but the transfer gates as well. Then it is possible to "select" the chip (to receive data, for instance) by energizing its input transfer gates. Sometimes two inputs are provided to energize the transfer gates, the action being that the gates will be activated when BOTH of those "chip select" lines are ON (at +5 volts). That's accomplished with an AND circuit, a basic building block the output of which operates only when both input A AND input B are ON.

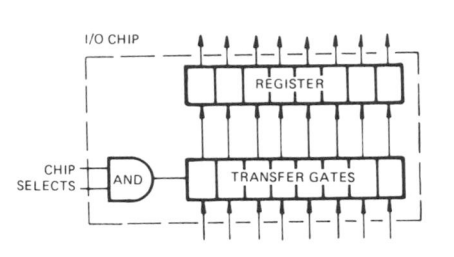

So we could use three I/O chips, connected to a common data bus coming out of the microprocessor to receive the twenty-one bits of data. That data will be emitted as three groups of seven bits each, one group after the other. As each group appears on the bus, the chip select lines will be used to direct the data on the bus to the correct I/O chip, which will hold it and direct it to the proper 7-segment display.

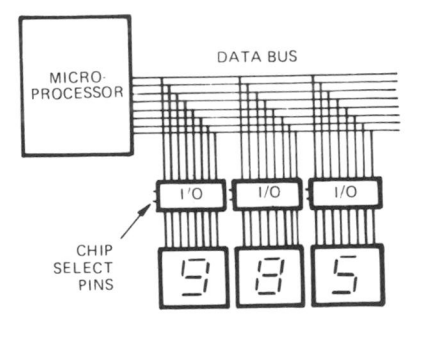

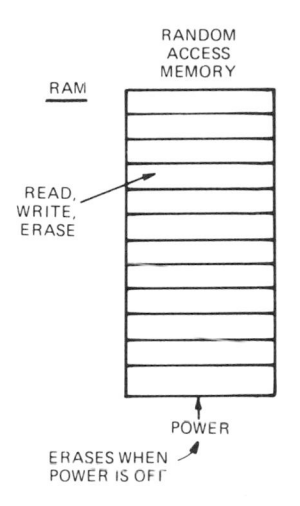

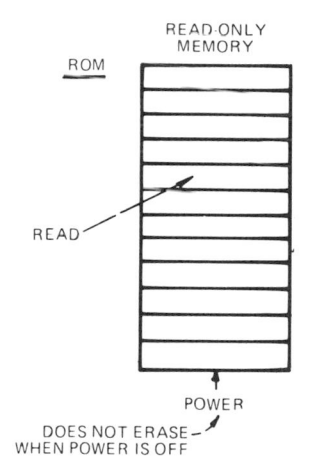

As we said, the table of values, in which we are going to store the gallon readings corresponding to each of the numbers from 0 to 200, we can store in a memory, which the microcomputer can read. These memories, or groups of registers, come in two basic types: random-access memories, or RAMs, which store data only as long as they are supplied with electric power, and which can be written into, read, rewritten, and erased (by rewriting); and read-only memories, or ROMs, which can store data even when they are turned off, but which normally cannot have their contents altered by the microprocessor during regular processing. Our table, once worked out, need never change, so a ROM would be the natural choice; then the data will not be lost when the microcomputer is turned off.

Actually, some in-between types of memory exist, which can be erased with special equipment and rewritten with special voltages. And there is a distinction between ROMs that are filled with their data at the factory, and those that can be programmed (but permanently) by the

user; the latter are often called
PROMs, or programmable read-
only memories.

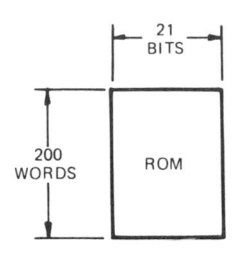

Our strategy here, then, will be
to store the table in a ROM; but
now we have to ask what the
best format is in which to store
it. One method is to store a full
21 bits for each of the numbers:
exactly the 21 bits we need to
send to the I/O chips and on to
the displays. We would be storing
the number of gallons in a special form of decimal numbers,
where each digit is represented
by 7 bits — just the 7 bits needed
to drive the 7-segment display.

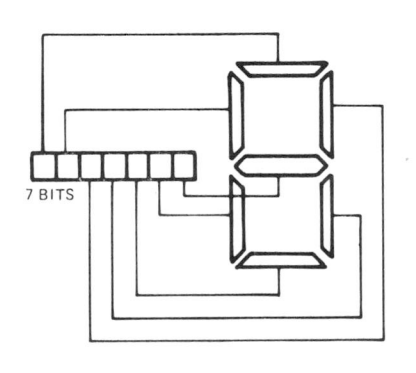

Another method would be to try
to use the fewest possible number of bits to represent the number of gallons. To do that, we
would carry the 200 values as
full binary numbers. We're measuring to the nearest one-half gallon, so we'd choose the least significant bit to represent half a
gallon. Then the next bit to the
left would represent one gallon;
the next bit, two gallons; then 4,
8, 16, 32, and 64. That makes 8
bits, which is, of course, ade-
quate to express 256 possibilities
— and we have only 200 possibil-

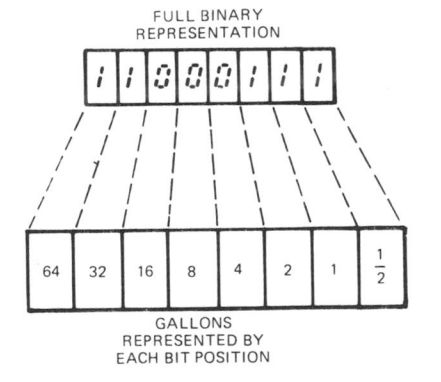

FULL BINARY
REPRESENTATION

| 64 | 32 | 16 | 8 | 4 | 2 | 1 | $\frac{1}{2}$ |

GALLONS
REPRESENTED BY
EACH BIT POSITION

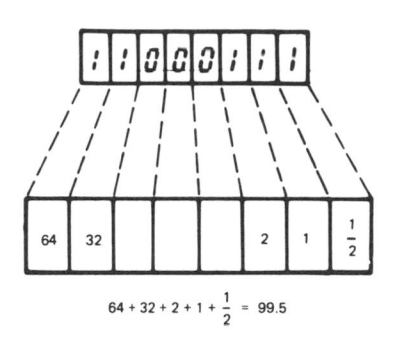

$$64 + 32 + 2 + 1 + \frac{1}{2} = 99.5$$

MEMORY

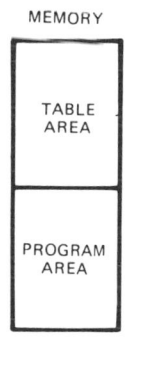

BINARY-CODED-DECIMAL

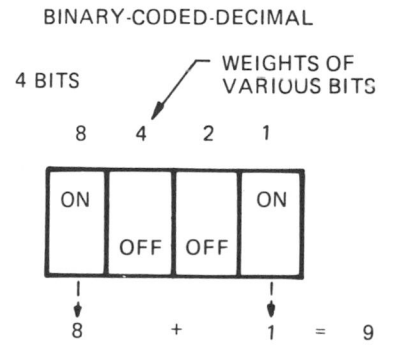

ities to express. The largest binary number we would have to use would be 11000111, which is decimal 199, corresponding to our maximum display of 99.5 gallons.

We could use such a system. However, we then have to include a section of computer program dedicated to converting this binary number back to a form that can be sent out to the displays, and this conversion program also consumes memory. Memory was exactly what we wanted to save when we chose the full binary representation — so we see that the most highly "efficient" representation may not be the most effective overall solution to the problem.

Still another way to store the table would be to store decimal digits, but store each digit in the fewest possible bits, that is, four (in a code where the individual bits represent 8, 4, 2, and 1). This is called "binary coded decimal," or BCD, and we would not need more than 12 bits to store the three decimal digits we want. There could then be a small second table, consisting of ten entries of 7 bits each, in

which the 7-segment representation of a digit could be looked up. Again, however, the second lookup process requires more computer program storage.

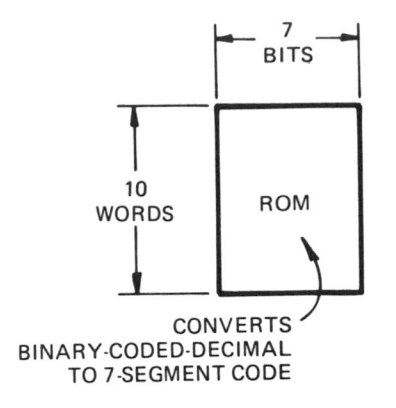

CONVERTS
BINARY-CODED-DECIMAL
TO 7-SEGMENT CODE

All things considered, especially the fact that the original table of all 21 bits only needed to be 200 entries long, the best conclusion probably is to use a single lookup on that one larger table. At least the program will be quite a bit shorter, and the program development and debugging time — always a major consideration in any computer application — will certainly be less. In a real application, the number of tank-measuring systems to be built and sold would be taken into consideration, as would the size of the table. A larger table, and a larger number of systems to be built, might warrant the extra memory space for computer programs and the extra engineering time to develop and debug the programs (and the extra risk that the more complex program will have an unseen "bug" that will require recall of many production units).

5

A BLOCK DIAGRAM AND
A SOFTWARE DIAGRAM

Let's examine now the data processing steps that the microcomputer will have to go through, but not in the form of detailed computer instructions; instead, let's first examine the steps at the highest level, that of the application itself.

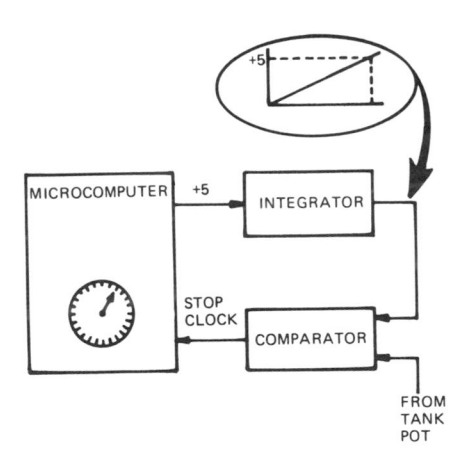

As we have said, the microcomputer places a steady +5 volts on the integrator's input, then acts as a clock, counting 200-microsecond periods by increasing the value in a register. When the comparator "fires," the register is consulted and its contents used as the entry into a table of 200 entries, each of which has the entire 21 bits we need to send to the displays. Then the whole process is repeated.

Now all this activity can be expressed by a combination of a

block diagram and a special software diagram. The block diagram should show the important parts of the system, and also carry notations about the transfer of information, and the conditions under which important signals take place.

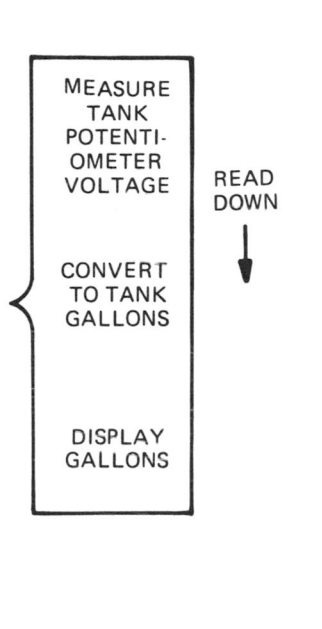

The software diagram, used to explain the operation of the computer program, requires a little more explanation. It is an adaptation of a chart that is called a Warnier-Orr diagram after its originators. It is made up of boxes representing different parts of the program; within each box, operations to be performed are listed from top to bottom.

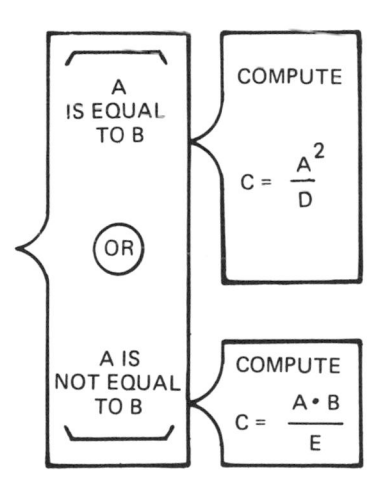

Sometimes, however, there is a choice of one of two actions to be taken, depending on some condition. This is shown by listing the two conditions within brackets above and below, and with an "OR" symbol between them. It's understood that the two conditions are mutually exclusive; only one of them can hold. Opposite each condition, there will be another box, with a bracket point on its left side, which points at the condition.

POT VOLTAGE IS A (NON-LINEAR) MEASURE OF TANK GALLONS.
RAMP VOLTAGE <u>WHEN</u> <u>COMPARATOR</u> <u>FIRES</u> IS A MEASURE OF POT VOLTAGE.
TIME IS A MEASURE OF RAMP VOLTAGE.
TIMING COUNT IN MICROCOMPUTER IS A MEASURE OF TIME.
LOOKUP TABLE IN ROM CONVERTS TIMING COUNT DIRECTLY TO GALLONS.

START RAMP AND TIMING COUNT TOGETHER
WHEN RAMP = POT, CONVERT TIMING COUNT
TO GALLONS BY TABLE LOOKUP.

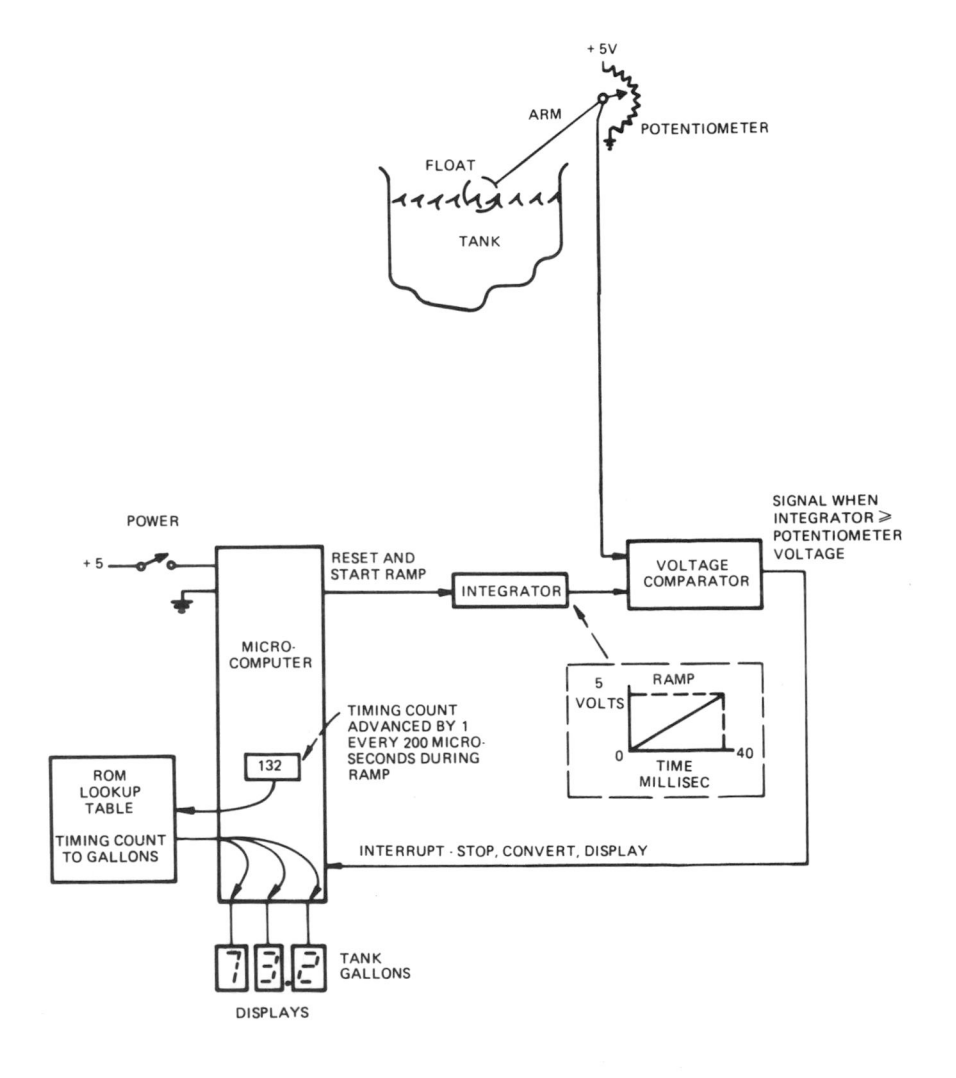

In it will be the actions that are to be carried out if that condition holds.

It's understood that the sequence of actions down the box on the left is momentarily suspended while the actions in the box opposite the condition are carried out. Then the sequence down the left box is resumed.

STARTING
POINT

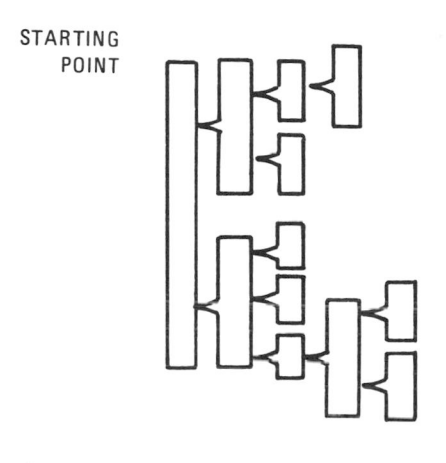

Of course, it's possible for the boxes on the right to have conditional statements in them, too, and to lead to boxes still further off to the right. This can be carried for many levels if necessary; large programs often have ten or more such levels.

Sometimes it is desirable to have a sequence of actions repeated several times. In that case, the computer has to be told when to stop; this is noted on the diagram with the same sort of a pair of conditions. Again, there is a pair of brackets above and below, and an "OR" symbol separating the two conditions, which must be mutually exclusive. However, above the upper condition is the notation "REPEAT WHILE."

It's understood that the box to the right of the upper condition

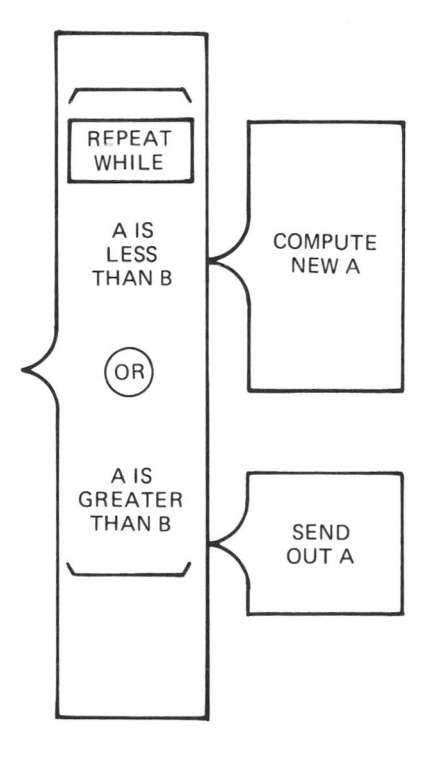

will be repeated as many times as necessary, until the upper condition no longer holds. Then the box opposite the lower condition is carried out.

The clearest method by which to use these diagrams is to show them developing level by level. We begin here with just the first two levels: beginning at the top of the leftmost box, we read that the "power comes on." As noted, this happens only once. At that time, any setup operations that must be done when the power comes on are carried out.

Then, while the "power is still on," the system will repeat this sequence: measure tank potentiometer voltage; convert to tank gallons; display those gallons.

However, when the power goes off, then the entire process stops.

Now the process can be detailed to the next level down; in doing so, the summary of actions we had in the right box from the previous diagram — "measure potentiometer voltage, convert to tank gallons, display gallons" — is replaced by a breakdown showing more exactly what happens. First

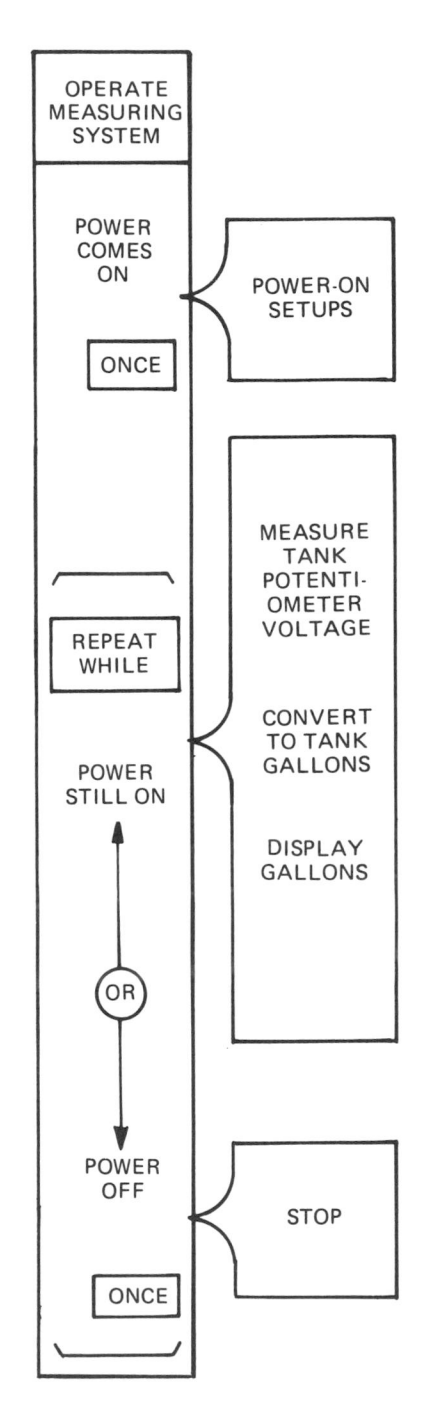

there is executed something labelled "begin measurements": the timing count (in the register in the microcomputer) is set to zero, and the integrator voltage ramp is started (from zero volts.)

After that, there is a "REPEAT WHILE": as long as the ramp voltage is less than the potentiometer voltage, so that there is NO COMPARATOR SIGNAL, the system will advance the timing count once every 200 microseconds — that is, count up the number in the register.

But when the ramp voltage reaches the potentiometer voltage, and the COMPARATOR SIGNAL fires, then the lower sequence takes place: the integrator is shut off; the main timing count (the number in the register) is used to look up the digits (representing the gallons); and the digits are sent to the displays.

Now, all of this can be shown to at least one more level of detail. We'll save that diagram, however, for later, after some details of the operation of the microcomputer become clear.

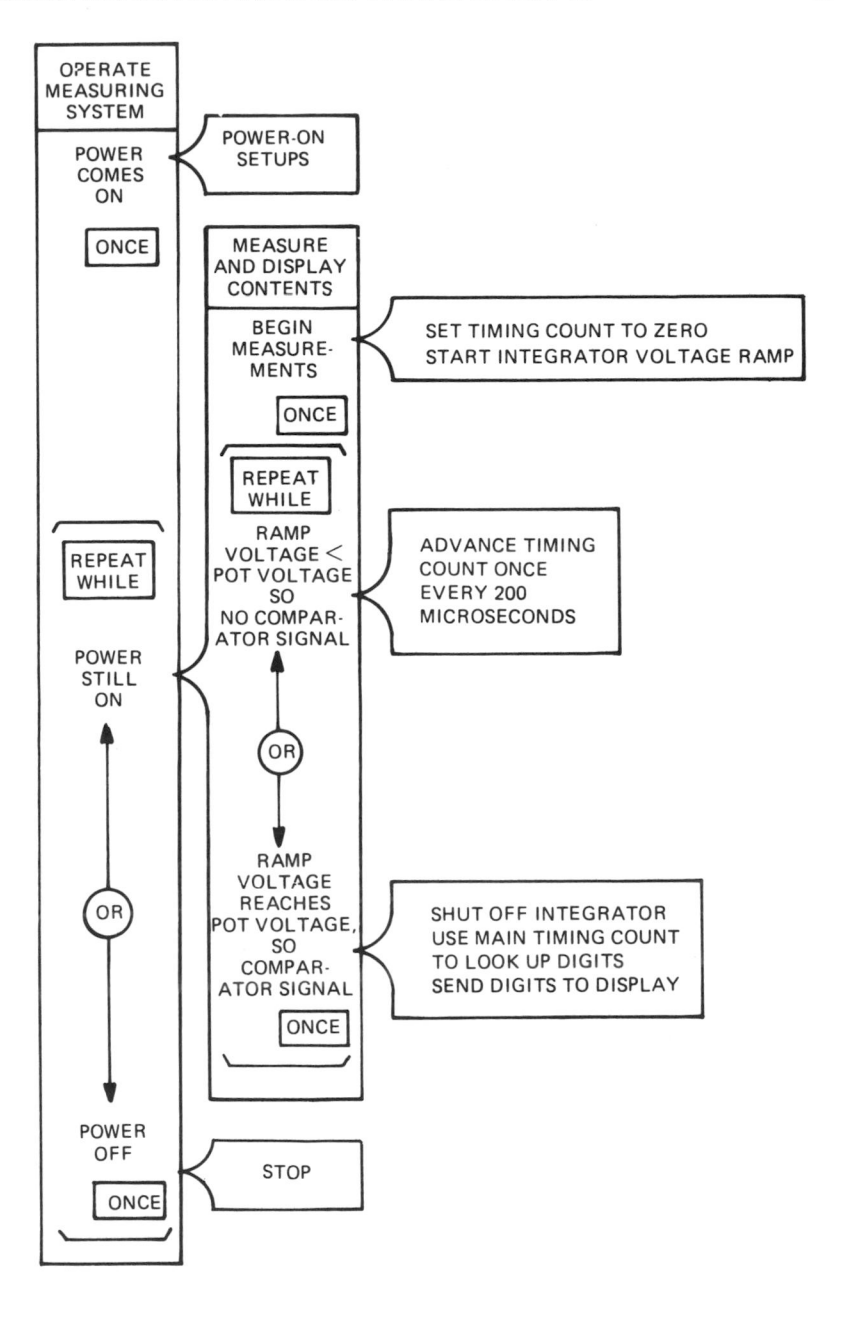

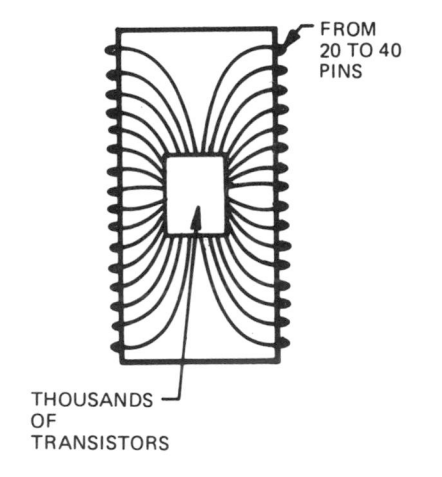

FROM 20 TO 40 PINS

THOUSANDS OF TRANSISTORS

When we come to implement all of this in an actual microcomputer, built of real chips available on the market, we find that many small peculiarities of a particular manufacturer's chips have to taken into account. That, of course, is life in the electronics industry; each manufacturer tries to put into his chips what he believes will be the most useful for the cost involved, and he is fighting, most of all, the fact that practical packaging of chips limits the number of input/output pins that can be on each chip. Internal circuitry of the chip, however, is now so inexpensive that all sorts of complexity and trickery can be used to maximize the usefulness of the chip for a given number of input/output pins. Hence the often very peculiar and seemingly over-complex functions inside the chip.

We've selected, as the micro-processor around which to build our microcomputer, one of the most popular 8-bit processors: the Motorola MC6800. Let's look now at just those aspects of this microprocessor that we will need to know about, to apply it to our problem.

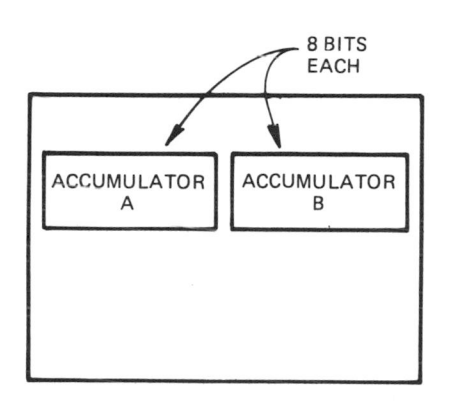

8 BITS
EACH

ACCUMULATOR
A

ACCUMULATOR
B

Inside the MC6800 are six registers with which the user must concern himself. First of all, there are two "accumulators," or registers in which to manipulate data. These are both eight bits long, and are called Accumulator A and Accumulator B. Many of the 72 different instructions we can give the microprocessor cause it to manipulate data in these accumulators — for instance, adding a number in memory to a number in one of the accumulators, or incrementing a number in one of the accumulators by 1. Some instructions tell the microprocessor to consider these two accumulators as one long 16-bit register.

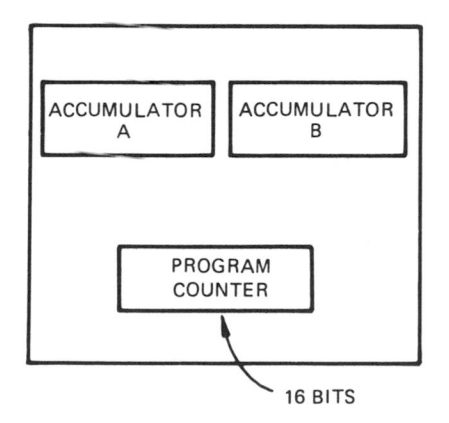

ACCUMULATOR
A

ACCUMULATOR
B

PROGRAM
COUNTER

16 BITS

Then there is a 16-bit register that holds the address, in memory, of the instruction to be executed next. When it is time to "fetch" the next instruction

from memory, whatever is in this register is what is presented to the memory, along with a command to read; the answer will be an eight-bit "byte" from the memory, which will be brought into the microprocessor. That byte will be the instruction, or perhaps part of the instruction, since some instructions require two or three bytes. This information goes into other registers inside the microprocessor; but those are not registers with which the user must concern himself, so they are not part of the ones we are describing.

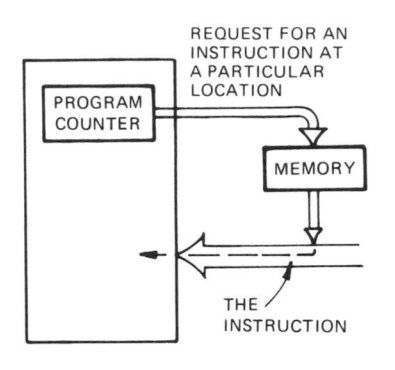

The reason that the program counter is of concern to the user is that its contents can be changed, saved, or replaced by the user by using instructions. Doing those things will cause the microprocessor to begin executing instructions in some part of the memory apart from where it has been executing them, so the user has control, through instructions, over what the processor will do next. It is very convenient to think of the contents of the program counter as a "pointer," or arrow, that points to the particular location in memory. This is much more meaningful than

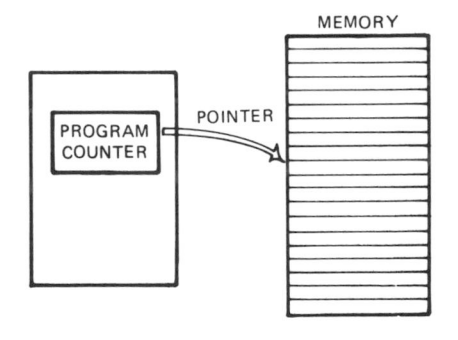

thinking of it as just a 16-bit binary number.

The functions of the other three registers in the microprocessor (that the user must consider) are somewhat specialized, and we will describe them separately as the occasion arises in discussing our example.

As we said before, the microprocessor has eight pins to which wires can be attached for the movement of data into and out of the unit, generally to and from one of the accumulators, though some instructions bring other registers into play. These eight wires are called the data bus, and they are special in that information can flow in both directions along the bus (at different times). This is in contrast to most wiring in logic systems, which carry data in one direction only.

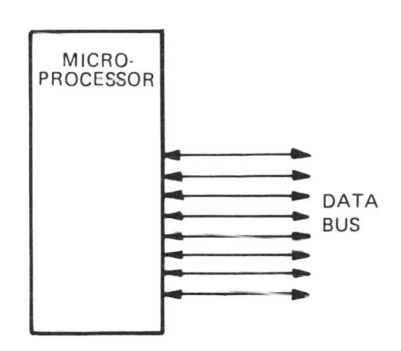

The microprocessor also has a set of 16 pins, to which wires can be attached, which are meant to carry binary numbers called addresses; together they are called the "address bus." The address bus carries information outward only, from the microprocessor to

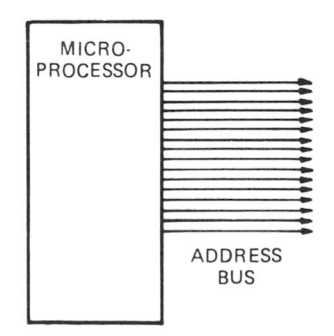

ROMs, RAMs, and I/O chips. The signals on the address bus are used both to select a certain memory or I/O chip, and to select a particular location or register inside that chip. We will spend some considerable time dwelling on how this is done.

Finally, there is a collection of assorted signals that enter and leave the microprocessor. Some of these carry control signals back and forth between the microprocessor and the ROMs, RAMs, and I/O chips, so these are grouped together and called the "control bus." The remaining wires go back and forth between the microprocessor and some support chips, and these are generally of no concern to the user except that he has to wire them up according to some directions. Some day, support chips will not be necessary, as all the needed functions are brought onto the microprocessor chip itself. Then all of the wiring, except the three buses and the power connections, will disappear inside the microprocessor.

A few of these special connections are worth mentioning at this point:

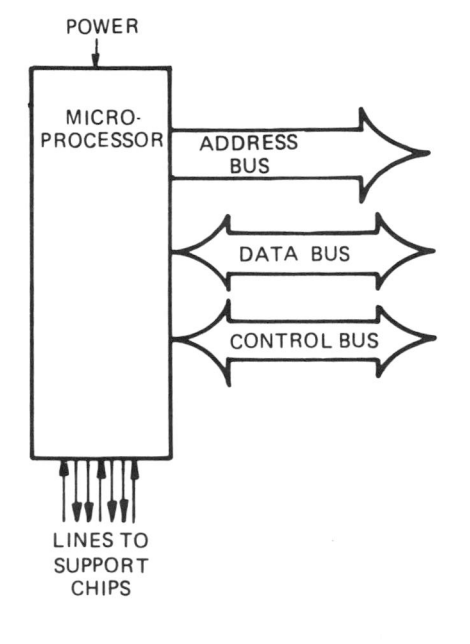

The power supply needs of the MC6800 are +5 volts and ground. That's quite modest compared with some other microprocessors, which require multiple power supply voltages.

The MC6800 also requires two "clock" signals, which are square waves (signals which alternate between +5 volts and zero volts, with sharp rises and falls). These signals are called the "phase 1 clock" and the "phase 2 clock," and they differ only in that phase 1 is at +5 volts while phase 2 is zero volts, and vice versa. In our application each clock signal will stay at +5 volts for .45 microsecond. With that timing on the clocks, a single machine cycle, as it's called, will last one microsecond, and the times of instructions are then known because they each take a known number of machine cycles (for instance, a jump instruction — that is, an unconditional branch — takes three machine cycles, so it will take three microseconds in our example). You can operate with a slower clock if your application demands it.

The MC6800 also has a RESET line, which in normal operation is fed with +5 volts; a momentary pulse at zero volts will reset

the processor. After such a reset, the microprocessor will begin executing instructions from an address you can specify, because it looks in a predetermined place in memory where you have put that address. In the MC6800, your starting address is always stored at addresses 1111 1111 1111 1110 (decimal 65,534) and 1111 1111 1111 1111 (decimal 65,535), that is, at the highest possible addresses; and it is up to you to see that the new starting address is located there.

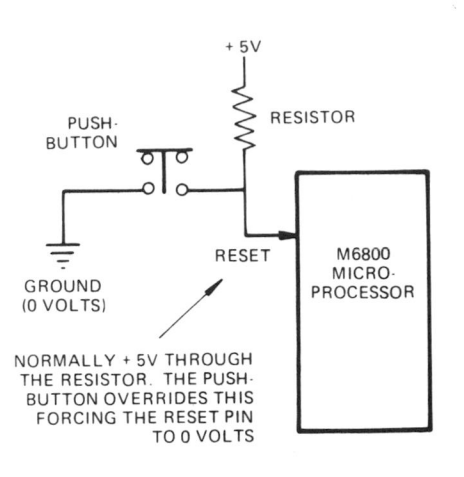

PUSH-BUTTON

RESISTOR

+ 5V

RESET

M6800 MICRO-PROCESSOR

GROUND (0 VOLTS)

NORMALLY + 5V THROUGH THE RESISTOR. THE PUSH-BUTTON OVERRIDES THIS FORCING THE RESET PIN TO 0 VOLTS

INSIDE THE
MICROPROCESSOR

Now let's consider the specific instruction sequence we need to make the microcomputer operate as we have planned. Take, first of all, the incrementing of the register that will hold the number representing the integrator's time: how is it done? What register in the microprocessor is used, and what instruction is used to add 1 to it? And — how do you, as the programmer, write that instruction down, and then have it converted into the form in which the microcomputer needs it?

We have already described accumulator A and accumulator B. Among the many instructions that involve these registers, there are two called "increment A" and "increment B" — so only a single instruction is needed to add 1 to either of these registers.

MICROPROCESSOR M6800

ACCUMULATOR
A

ARITHMETIC
UNIT

ACCUMULATOR
B

REGISTERS,
EACH 8 BITS LONG

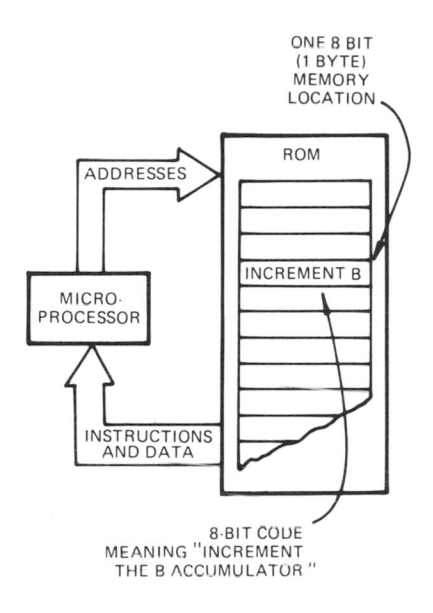

ONE 8 BIT
(1 BYTE)
MEMORY
LOCATION

ADDRESSES

ROM

INCREMENT B

MICRO-
PROCESSOR

INSTRUCTIONS
AND DATA

8-BIT CODE
MEANING "INCREMENT
THE B ACCUMULATOR"

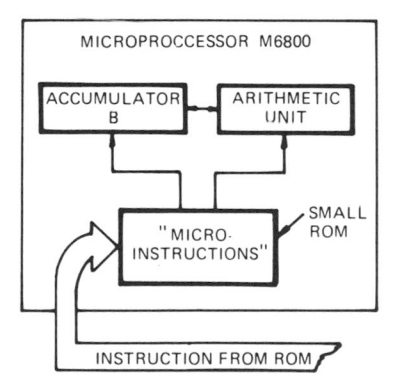

MICROPROCCESSOR M6800

ACCUMULATOR
B

ARITHMETIC
UNIT

"MICRO-
INSTRUCTIONS"

SMALL
ROM

INSTRUCTION FROM ROM

This single instruction will "reside" in the read-only memory, or ROM, and will be executed when the microprocessor reads it from the ROM, as part of the entire sequence of instructions. This particular instruction is just eight bits, or one byte, long, and so occupies a single memory location in the ROM. When it is read by the microprocessor, it is used as the signal for the microprocessor to go to a particular part of a smaller ROM that is right inside the processor chip, and which contains "microinstructions" controlling what is to be done. In this case, the contents of the A accumulator (or the B accumulator) are fed into the arithmetic unit of the processor, along with a "1," and the arithmetic unit is instructed to add them. The answer is then placed in the appropriate accumulator. To the programmer, however, all this is invisible. All he needs to know is that, for instance, he wants the B accumulator incremented. What is especially useful is that he does not need to know even the eight-bit code for the instruction. He actually writes just "INC B"; a computer program, known as an assembler, will then convert "INC B" to

"01011100," the code for INC B in the M6800 microprocessor.

The assembler computer program can be run on any of a number of different types or sizes of computers. It can, for instance, be run on a large computer in a data center, using punched cards as the input medium. It can also be run on a large computer in a central location, accessed by telephone from a terminal, a typewriter-like device (or sometimes a keyboard and a video screen). The assembler program can also be run on a minicomputer or even a microcomputer, even one exactly like the one for which the program is being written — if it is equipped with all the necessary "peripherals," such as a keyboard and video screen, floppy disk unit, and a paper tape punch, or just a teletypewriter and a cassette recorder.

We know, then, that our program will at least contain the instruction "INC B," since we have elected to use accumulator B to hold the number that will be the measure of the integrator time. Suppose we just repeated that instruction over and over, until the comparator says that we have

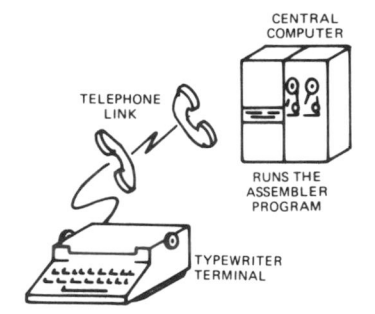

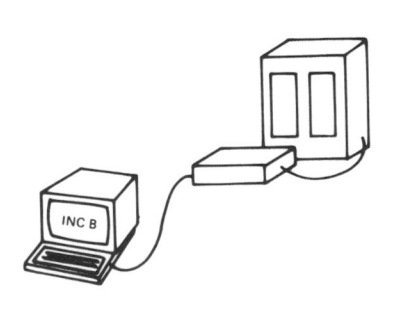

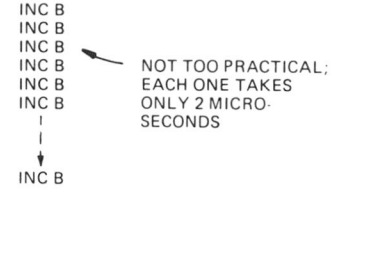

INC B
INC B
INC B
INC B   NOT TOO PRACTICAL;
INC B   EACH ONE TAKES
INC B   ONLY 2 MICRO-
      SECONDS
INC B

reached the pot voltage? Unfortunately, that will not be quite practical. The "INC B" instruction takes just two microseconds to execute, if the clock frequency is set so that each cycle of the clock is one microsecond. If we set the integrating time so that the integrator reaches full scale when the B register reaches a count of 200, that would mean that the integrating time would have to be 2 x 200 = 400 microseconds, which is rather too fast for a practical integrator. What we would like is to have the integrator finish a complete ramp (corresponding to a full tank) in about 100 times that figure, or about 40 milliseconds or so. That's a twenty-fifth of a second, quite fast enough to make the displays very responsive (in human terms) to any changes in the liquid in the tank.

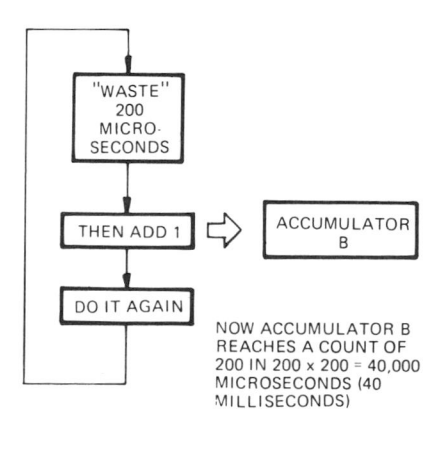

"WASTE"
200
MICRO-
SECONDS

THEN ADD 1 ⟹ ACCUMULATOR B

DO IT AGAIN

NOW ACCUMULATOR B
REACHES A COUNT OF
200 IN 200 x 200 = 40,000
MICROSECONDS (40
MILLISECONDS)

We have to have, then, some way to "waste time" between the times when we want to add 1 to the B accumulator. In fact, we would like to waste about 200 microseconds or so. In "wasting" it, though, we want to be sure that we are also measuring that time accurately.

The time-honored method for doing this is to place, in another register, some number, and then keep decrementing it by 1 (with a series of instructions that take a known length of time), checking all the while to see if it is zero yet. In our case we can, for instance, place the number 17 in a register, check it for zero, decrement it by 1, and loop back to check it again.

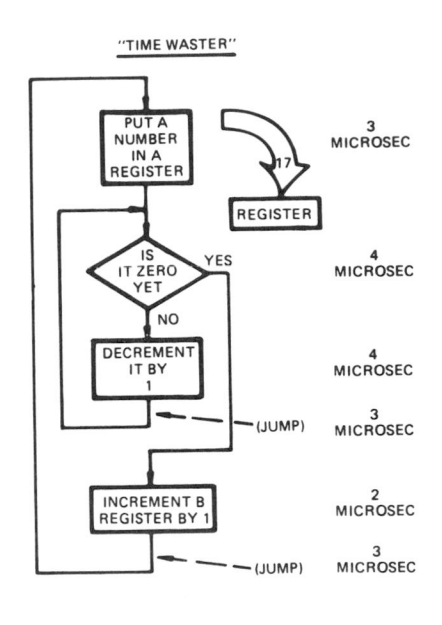

"TIME WASTER"

In the MC6800, placing the number in the register requires 3 microseconds (if our basic clock period is one microsecond, as we've chosen). Checking for zero (and branching if it is) takes 4 microseconds. Decrementing the register by 1 takes 4 microseconds, and jumping to another point in the program — shown on a flow chart just by an arrow — takes 3 microseconds.

Finally, incrementing the B register takes 2 microseconds, and a final jump takes another 3 microseconds.

Now we can divide the whole process into 18 "passes," as the chart shows: a first pass that includes setting the number "17"

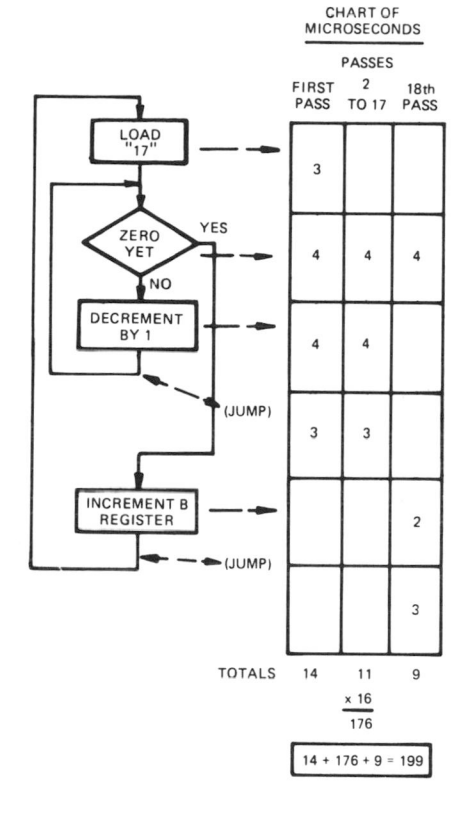

CHART OF
MICROSECONDS

PASSES

	FIRST PASS	2 TO 17	18th PASS
LOAD "17"	3		
ZERO YET	4	4	4
DECREMENT BY 1	4	4	
(JUMP)	3	3	
INCREMENT B REGISTER			2
(JUMP)			3
TOTALS	14	11	9

x 16
176

14 + 176 + 9 = 199

into the register; then 16 more, as the register is decremented down to zero one step at a time; and finally the 18th pass, when the zero check comes out "yes," and the path leads us down to increment the B register and then jump all the way back to the top to load the register with "17" again.

The first pass will take 14 microseconds; passes 2 through 17 will take 11 microseconds each, and since there are 16 of them they will consume 176 microseconds; and the last pass will take 9 microseconds. All that adds up to 199 microseconds, acceptably close to our goal of 200.

We can compensate for the fact that it has turned out to be 199 microseconds instead of 200 by altering the slope of the ramp, so that it takes (200 counts x 199 microseconds) = 39,800 microseconds = 39.8 milliseconds (instead of the 40 we originally intended) to reach +5 volts.

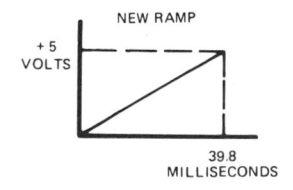

NEW RAMP

+5 VOLTS

39.8 MILLISECONDS

Notice that we very naturally slipped in the notion that, before decrementing the register by 1, we would "check it" for zero, doing one thing if it is not zero

and another thing if it is. That notion — of the conditional branch, or conditional jump — is at the heart of the programmable computer. It is its single most important property, the capability that separates a computer from a mere controller of sequence, like a drum controller for traffic lights. More than anything else, it is the conditional jump that makes a computer able to mimic the intelligence of life: its ability to take different courses of action depending on the outcome of previous calculations.

The register we will choose to hold the "17" could be the A accumulator. However, to illustrate that other registers exist in the machine and can be used, we will use what is called the index register. Its regular use in the machine is for something else — which we will presently come across — but it, too, can be incremented and decremented by a single instruction (here the decrement instruction is "DEX"). The index register is 16 bits long, which means that it can count to 65,535 if we needed it (here we do not, since 17 is as high as we need).

The instruction which will load the number "17" into the index

MICROPROCESSOR M6800

ACCUMULATOR A

ACCUMULATOR B

ARITHMETIC UNIT

INDEX REGISTER

16 BITS LONG
CAN BE INCREMENTED
AND DECREMENTED

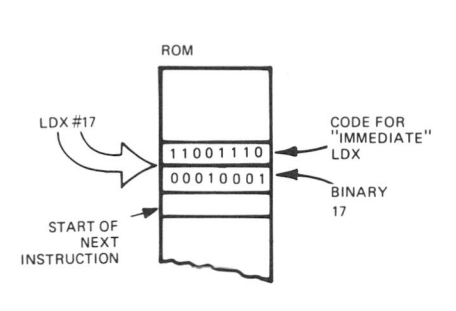

register is "LDX #17." LDX stands for "load index register," as we might suspect, and "#" tells the assembler program that we want the "17" put into the ROM right in the stream of instructions, as part of the LDX instruction, and that we want a particular eight-bit code for the "LDX" itself that tells the microprocessor that it's to pick up the "17" at that place. Packaging a constant into the instruction that way is called the "immediate" form of the instruction; it's available for only certain instructions. We would read the notation "LDX #17" as "load index register immediate with 17."

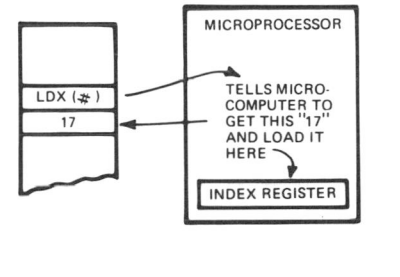

Having loaded 17 into the index register, we now have to test it for zero, and, if the answer is that the index register hasn't reached zero yet, decrement it with the DEX instruction, and loop back to test it again.

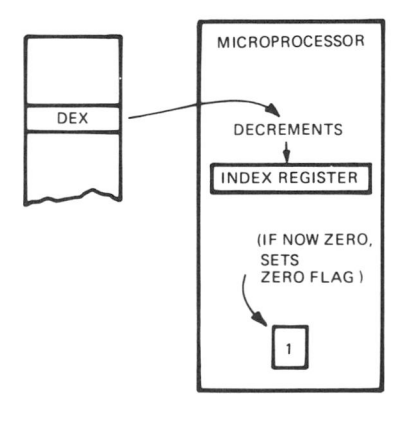

The DEX instruction, like many others, has a built-in test for zero. If zero is reached when the decrementing is done, then a bit inside the microprocessor, called the zero flag, is set. The condition of

that flag can then be examined in a "branch" instruction (of which a number of types are available), causing the microprocessor to jump back (or forward) to another address if the desired condition was satisfied. The LDX instruction automatically resets this flag, so it won't be set the first time we test it. In our case, the BEQ instruction is what we want: "branch if equal to zero," that is, branch if the zero flag is set. If the branch is done, it takes us down to the "INC B" step; if it's not done, the program simply advances to the DEX step, and then we want it to jump back to test for zero again.

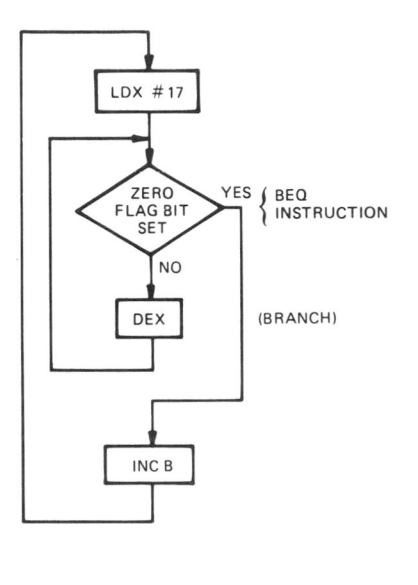

That jump can be done with a JMP instruction. The jump from the INC B back to the top can also be done with a JMP.

Now, however, we are in a quandary. We know that the series of instructions

        LDX #17
        BEQ
        DEX
        JMP
        INC B
        JMP

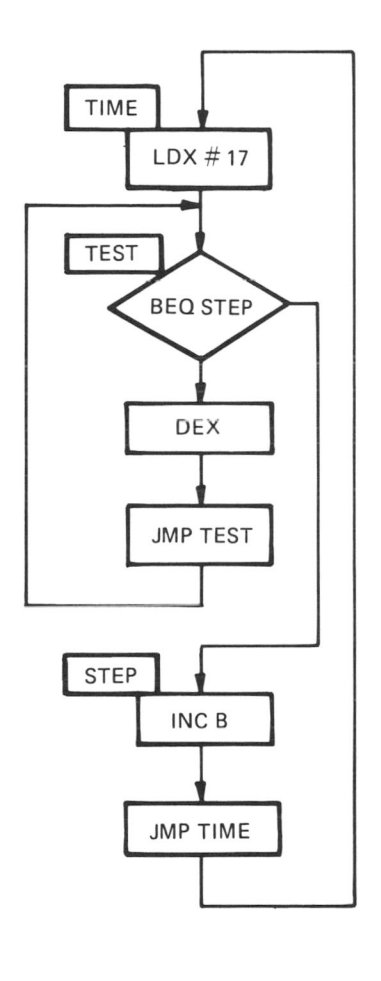

is going to lie in sequence somewhere in the ROM, but we don't
know yet the exact addresses.
Still, we have to indicate the
place. Here again, the assembler
program has facilities to help us.
For instance, we can assign
"labels"—combinations of up to
six letters or letters and numer-
als—to various instructions, and
then indicate in the jump and
branch instructions what label is
to be jumped or branched to. So
we can put the three labels TIME,
TEST, and STEP on the proper
instruction, and now the whole
series will read this way:

> TIME   LDX #17
> TEST   BEQ STEP
>   DEX
>   JMP TEST
> STEP   INC B
>   JMP TIME

When the assembler program has
finished assigning all the instructions to definite addresses in the
ROM, it will note the addresses
of the instructions marked with
the different labels, and arrange
things so that the jump or branch
instructions will cause the program to go to the proper memory address for the next instruction.

What actually happens in the microprocessor in a branch or jump instruction? Inside the processor itself is a special register known as the program counter, which keeps track of the place in memory from which the current instruction is being read. Ordinarily, instructions are directly in sequence in memory, so all the program counter has to do to step through the instructions is to keep incrementing itself by 1. Sometimes, when an instruction occupies more than one memory location (like the LDX #17, which occupies two locations counting the "17"), the program counter is told to increment itself more than once in a given instruction, as the various items are read. The program counter is 16 bits long, so it is possible for it to generate addresses for instructions that are in up to 65,536 locations.

When a branch or jump instruction comes along, however, the contents of the program counter can be changed suddenly to a value that is carried right along with the instruction. In a jump instruction, that can be 16 bits of address, going to anywhere in memory. In what is called (in the

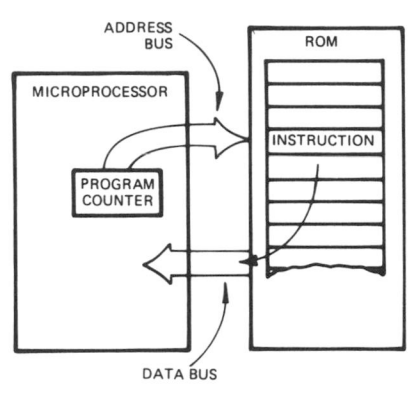

PROGRAM COUNTER ISSUES ADDRESS ON ADDRESS BUS TO ROM; INSTRUCTION AT THAT ADDRESS GOES INTO MICROPROCESSOR ON DATA BUS

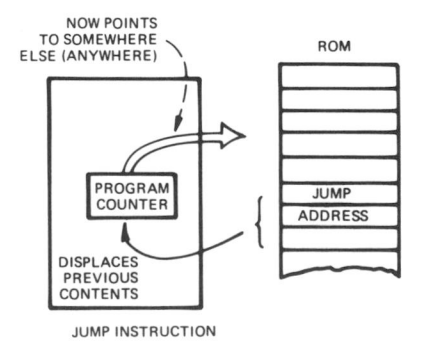

JUMP INSTRUCTION

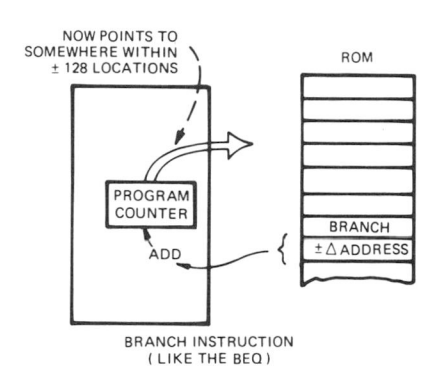

NOW POINTS TO
SOMEWHERE WITHIN
± 128 LOCATIONS

ROM

PROGRAM
COUNTER

BRANCH
± Δ ADDRESS

ADD

BRANCH INSTRUCTION
(LIKE THE BEQ)

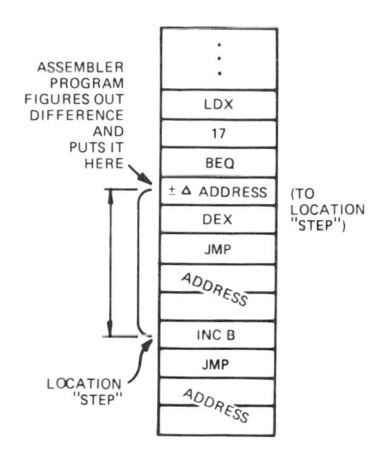

ASSEMBLER
PROGRAM
FIGURES OUT
DIFFERENCE
AND
PUTS IT
HERE

LDX
17
BEQ
± Δ ADDRESS
DEX
JMP
ADDRESS
INC B
JMP
ADDRESS

(TO
LOCATION
"STEP")

LOCATION
"STEP"

M6800) a branch instruction, however, it's limited to eight bits, and it doesn't mean a completely general address. It means, instead, the distance — as a number of locations, forward or backward, that we want to move. So it is the number that is to be added (algebraically) to the program counter. To figure this number out, we need to know, of course, exactly how many memory locations there are in all the instructions up or down to the target point. The assembler program relieves us of that tedious task; it knows that we want to go down to the instruction we marked "step," for instance, and it figures out what to put into the BEQ instruction to make that happen.

As we said, if the result of the test is that the index register just went to zero, we will branch down to the INC B instruction. That will be the instruction that increments the number we are holding in the B accumulator, since we have now timed out the 199 microseconds.

In this case, the "B" goes in a special column; this particular assembler program uses that column to identify whether the

instruction is to apply to the A or B accumulator.

Properly spaced out, then, as we would write it on a punched-card form, the timing loop looks like this:

```
TIME LDX   #17
TEST BEQ   STEP
     DEX
     JMP   TEST
STEP INC B
     JMP   TIME
```

The Warnier-Orr representation of this section of program shows that it is a REPEAT WHILE; that is, the path through the DEX, or decrementing, instruction is repeated while it has not yet been done 17 times. When it has, then the main timing count, in the B accumulator, is increased by 1, and the whole process is repeated.

Adding this section to the Warnier-Orr diagram we began in Chapter 5, we now have a more complete picture; yet, as we will see, it can be detailed still further.

That's the heart of the program, and it certainly looks small enough and easy enough. The rest of the program is not any more difficult to understand, but it entails considerably more writing.

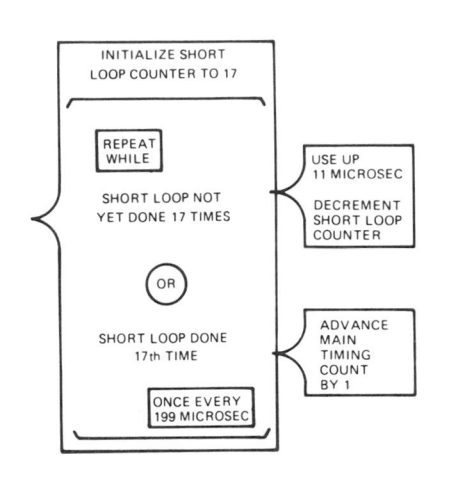

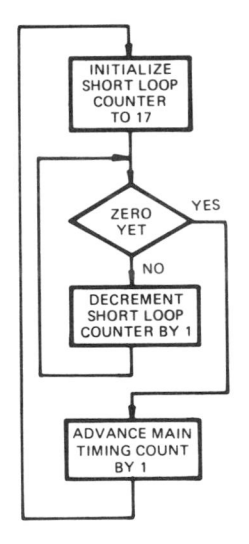

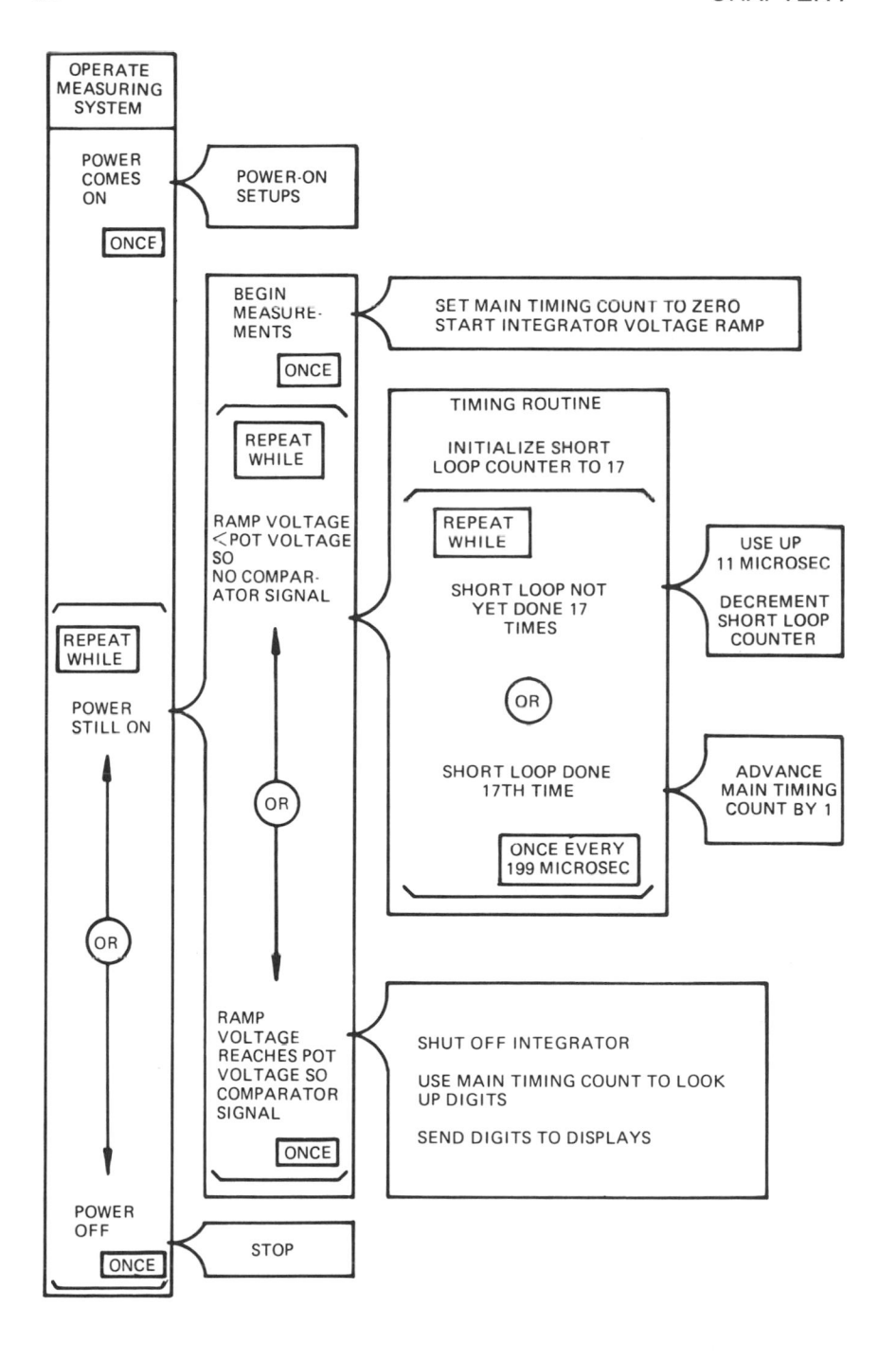

8 GETTING TO THE OUTSIDE: ADDRESSING

Understanding sections of code, we have seen, is not too difficult, certainly not when the functions of the instructions that are being used are fully understood. What most people find more involved is how the different devices external to the microprocessor are addressed: the RAMs, the ROMs, and the I/O chips that are the gateway to the rest of the world. There are ways to think about this, however, that make it straightforward.

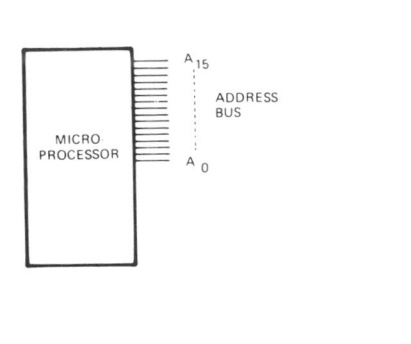

All of these external devices are selected by the microprocessor by placing addresses on the address bus, which goes to all those devices. The address bus, as we have noted, consists of 16 wires, or bits; we normally name them A_0, A_1, A_2, and so on up to A_{15}. A_{15} is the "most significant" bit of the address; that is,

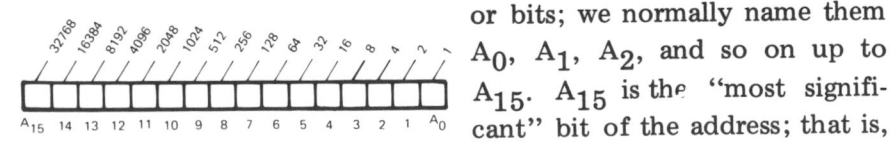

if we think of the address as a binary number that can run from 0 to 65,535, then A_{15} is the leftmost bit, the one that represents a weight of 32,768. A_0 is the least significant bit, the rightmost bit, the one that represents a weight of 1.

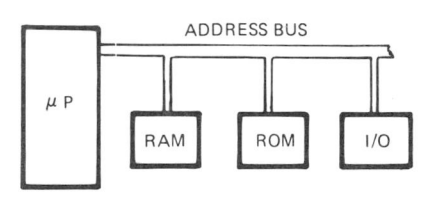

Binary numbers, however, especially when they are 16 bits long, are difficult to work with. It's certainly clumsy to use them in ordinary speech ("one oh oh one oh one one..."), and they are impossible to remember. On the other hand, it's also clumsy to think of the addresses in decimal form, because it takes a good deal of figuring to convert back and forth between the decimal number and the pattern of 16 bits. And it is necessary to work with the bit patterns in microcomputer work, because you are designing your own mix of peripheral chips (ROMs, RAMs, I/O chips), all with addresses peculiar to your application; so to know how to wire up their address lines, and how to debug the system, you have to be able to think comfortably about 16-bit addresses.

The compromise that is usually used is to work with the addresses in "hexadecimal" form, called "hex" for short. In this form, the 16 bits are broken up into four groups of four bits each; then a code, consisting of one letter or number, is used to represent each group. So an address that is 16 bits long can be given as four of these number codes: "FA34"; "05DC," and so on. The codes become quite easy to work with and remember. But best of all, it is very easy to convert them back and forth from the bit patterns.

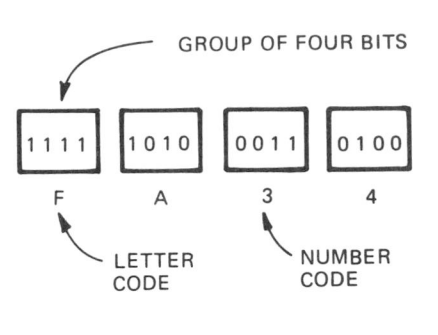

To do the conversion, the easiest way is to visualize certain things about the chart of the bit patterns versus the letter/number codes. Notice that the first four bit patterns begin with "00"; the next four with "01"; the next four with "10"; and the last four with "11." That is, of course, the standard sequence for a two-bit binary number: 00, 01, 10, 11.

Next, notice that the groups of four (going down the table) start with 0, 4, 8, and C.

Then, notice that the last two bits of each bit pattern go through the same sequence—00,

0	0	0	0	0
0	0	0	1	1
0	0	1	0	2
0	0	1	1	3
0	1	0	0	4
0	1	0	1	5
0	1	1	0	6
0	1	1	1	7
1	0	0	0	8
1	0	0	1	9
1	0	1	0	A
1	0	1	1	B
1	1	0	0	C
1	1	0	1	D
1	1	1	0	E
1	1	1	1	F

START WITH 00
START WITH 01
START WITH 10
START WITH 11

EACH GROUP
IS IN SEQUENCE

0	0	0	0	0
0	0	0	1	1
0	0	1	0	2
0	0	1	1	3
0	1	0	0	4
0	1	0	1	5
0	1	1	0	6
0	1	1	1	7
1	0	0	0	8
1	0	0	1	9
1	0	1	0	A
1	0	1	1	B
1	1	0	0	C
1	1	0	1	D
1	1	1	0	E
1	1	1	1	F

STARTS WITH 0

STARTS WITH 4

STARTS WITH 8

STARTS WITH C

01, 10, 11--as, for instance, you go through the set 4, 5, 6, 7.

All that's necessary, then, is to look at the first two bits to get the group starting point—0, 4, 8, or C—and then look at the last two bits to sequence yourself along from there. So 1011 would be in the group that starts with 8, and it would be the last one in the group, so it would be 8-9-10-11.

Going the other way, that is, starting with the letter/number code, first note the group it's in, and write the first two bits; then note where it is in the group, and write the last two bits.

Soon you'll find, of course, that you know all 16 codes by heart, going in both directions.

Once you can convert easily back and forth for individual letter/ number codes, you can easily convert from an address like "23EF" to its binary equivalent, "0010 0011 1110 1111." That's really the whole reason for using hex codes.

You may notice that hex codes are actually numbers in a number system having the base 16, and

that that is the derivation of the word hexadecimal. That's true, but it actually doesn't matter for our purpose here.

It's natural to think, however, that using "hex" addresses must mean a great many mental gymnastics, converting back and forth between hex and decimal in your head. Fortunately, that is almost never necessary.

The usual reason for wanting decimal numbers is to verify that you have enough room in a memory (one that you know is decimal 1,024 locations long). There is, happily, an easy way to think in hex, as it were, avoiding all the converting back and forth.

The key to this is to visualize a memory map. If the whole memory space is 65,536 locations, a convenient way to divide it up is with 256 rows of 256 locations each. Notice that a "1K" memory (1,024 locations) is now four rows of the map; a 4K memory (4,096 locations) is 16 rows.

If we have some way to "find ourselves" on the map, using the hex addresses, then the form of the map itself will tell how close

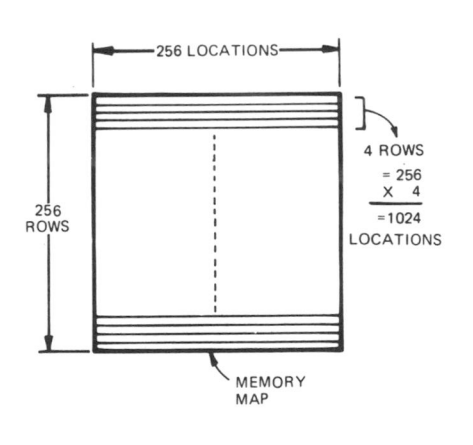

256 LOCATIONS

256 ROWS

4 ROWS
= 256
X   4
=1024
LOCATIONS

MEMORY MAP

we are to the end of a block of 1K or 4K or whatever. We can work very comfortably by moving around, mentally, inside the map.

The memory map can be marked off down the left side with the hex digits 0 to F. Notice that these will correspond to the leftmost digit of the address. In between each pair of digits there will be 16 rows of the map; these are numbered by the second digit. The map can also be marked off with digits 0 to F along the top edge; these correspond to the third digit. Between each pair will be 16 locations, and, of course, these are numbered by the fourth digit.

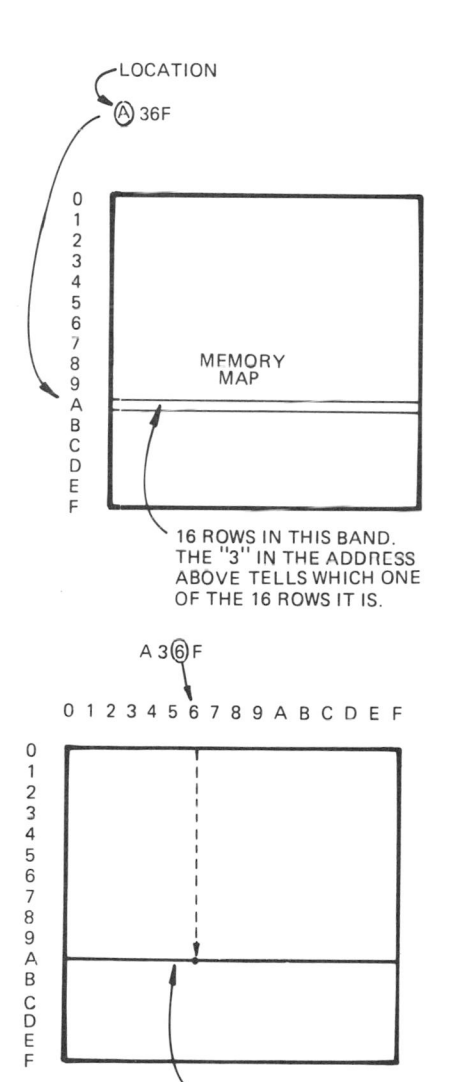

LOCATION

Ⓐ 36F

MEMORY
MAP

16 ROWS IN THIS BAND.
THE "3" IN THE ADDRESS
ABOVE TELLS WHICH ONE
OF THE 16 ROWS IT IS.

A 3 ⑥ F

0 1 2 3 4 5 6 7 8 9 A B C D E F

ROW "A3"

With this mental construction, then, it's possible to find your way around the map with the addresses, and the physical placement of locations on the map gives you an appreciation for the sizes of tables and programs. You know that to stay within a 1K memory (1,024 locations), you have to stay within four rows of the map. The need to think in decimal slowly goes away, for most purposes.

Visualizing the memory map in this way has another benefit: it fits in nicely with some test equipment that is available or can be constructed easily. Most of what are called "logic analyzers" are able to put, on an oscilloscope screen, a graph that shows what memory locations your microcomputer "visited," as bright spots on a memory map just like the one we have been describing. As your program runs, the spot on the oscilloscope describes a peculiar shape, and you can become quite familiar with the different parts of that shape and relate them to parts of your program. Debugging can be greatly aided, especially if facilities are also provided to move a cursor about the screen that will stop the operation when the microprocessor gets to that same point on the memory map. Suspicious places on the trace can then be examined in detail.

For those few times when converting from hex to decimal is unavoidable, it can be done with an ordinary calculator. "Hex" addresses have decimal weights associated with the different codes; in a group of four hex codes, the leftmost one has a

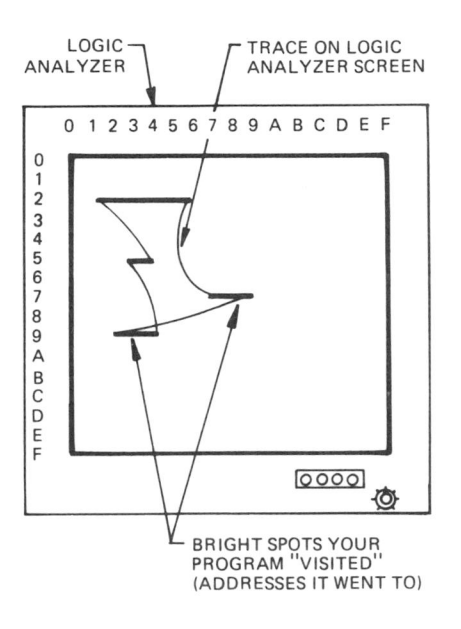

LOGIC ANALYZER

TRACE ON LOGIC ANALYZER SCREEN

0 1 2 3 4 5 6 7 8 9 A B C D E F

0 1 2 3 4 5 6 7 8 9 A B C D E F

BRIGHT SPOTS YOUR PROGRAM "VISITED" (ADDRESSES IT WENT TO)

weight of 4,096; the next, 256; the next, 16; and the rightmost, 1. The hex address "FA34," then, could be converted to decimal form by first noting that "F" is short for "15," so 4096 is multiplied by 15. "A" is short for "10" (decimal ten), so 256 is multiplied by 10. 16 is multiplied by 3, and 1 is multiplied by 4. The results are added together: $61,440 + 2,560 + 48 + 4 = 64,052$.

F	A	3	4
=15	=10	=3	=4
X 4096	X 256	X 16	X 1
=61440	=2560	=48	=4

64,052

The weights 4096, 256, 16, and 1, you may have noted, are 16^3, 16^2, 16^1, and 16^0.

CONNECTING THE ROMs AND RAMs

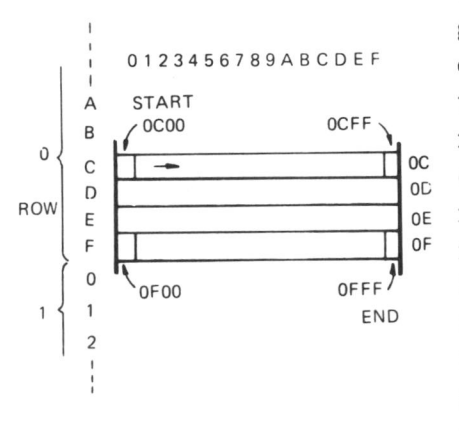

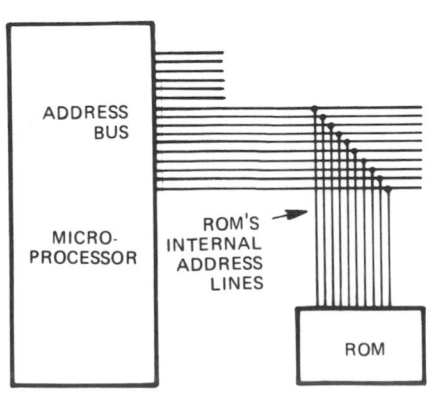

Suppose, now, that we are wiring up a ROM so that its addresses go from 0C00 to 0FFF (hex, of course). Remember that the first two hex digits in effect name a row of the memory map. Here, our ROM occupies the rows named 0C, 0D, 0E, and 0F. Does it occupy all of each one of the four rows? Yes; the numbers in the first row run from 0C00 to 0CFF; and the others run similarly, so that the fourth row runs from 0F00 to 0FFF. And, since each row contains 256 locations, the ROM must be 1,024 locations long.

To single out one of the 1,024 locations from the other 1,023, you need 10 bits of address, because it takes 10 bits to give that many combinations (2^{10} = 1,024). So, of the 16 address bits, the lowermost 10 (the right-

most 10) should be wired to the ROM's internal address lines. Once the whole ROM chip has been selected, those ten lines will tell it what location inside we want.

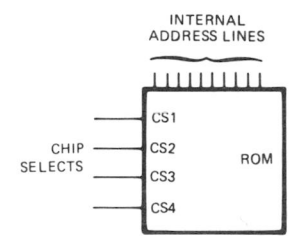

INTERNAL
ADDRESS LINES

CHIP
SELECTS

CS1
CS2　　ROM
CS3
CS4

But to select the ROM itself, the remaining six address lines (or at least some of them) have to be wired to the ROMs "chip select" inputs. These ROMs generally have four chip select inputs, which means that you might have to use some external logic elements (AND gates) to squeeze the six bits down to four. But that may not be necessary; it depends on how many different devices (ROMs, RAMs, and I/O chips) you are trying to distinguish. If you have only four chips to control, then you should need only two bits to get enough combinations to distinguish among them (00, 01, 10, 11).

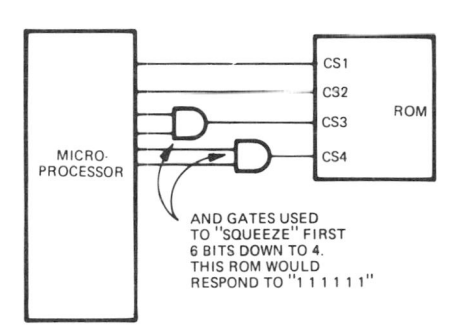

MICRO-
PROCESSOR

CS1
CS2
CS3　　ROM
CS4

AND GATES USED
TO "SQUEEZE" FIRST
6 BITS DOWN TO 4.
THIS ROM WOULD
RESPOND TO "1 1 1 1 1 1"

We can take advantage of that by choosing the addresses in the ROMs and RAMs properly. Since we are using the least significant ten address bits to call out the locations within the ROM, that is, address bits A0 to A9, we can use bits A10 and A11 to dis-

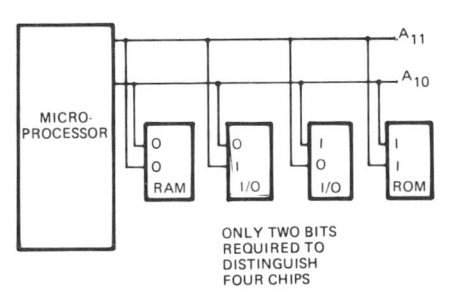

MICRO-
PROCESSOR

A_{11}
A_{10}

O
O
RAM

O
I
I/O

I
O
I/O

I
I
ROM

ONLY TWO BITS
REQUIRED TO
DISTINGUISH
FOUR CHIPS

tinguish among the various chips:
in our case, one ROM, one RAM,
and two I/O chips.

To do that, though, we have to
assign addresses so that bits A11
and A10 (reading from left to
right in the address) will be (for
instance) 00 in the RAM, 01 in
one I/O chip, 10 in the other I/O
chip, and 11 in the ROM (or
some such arrangement; it will
be clear later why we pick
those).

If we let bits A11 and A10 both
be 1's when we want the ROM
selected, then the address of the
starting location in the ROM
must be 0000 1100 0000 0000.
Converting each set of four bits
separately to hex, that comes out
0C00. Then, because the ROM is
1,024 locations long, the addresses in the ROM will run up
to 0FFF, which translates to
0000 1111 1111 1111. Notice
that the least significant ten bits
have gone from all 0's to all 1's,
which is, of course, exactly what
they should do to cover 1,024
locations.

With this arrangement, we want
the ROM to be selected when
bits A11 and A10 of the address

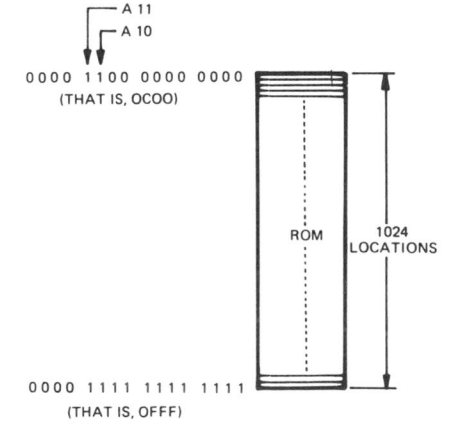

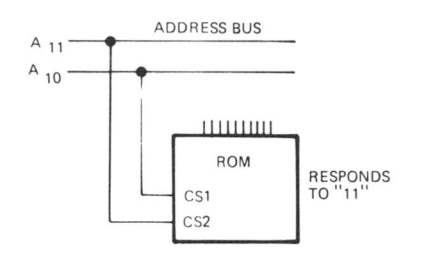

ACTIVE HIGH CHIP
SELECT INPUTS ··RESPOND
TO + 5 VOLTS, THAT IS, ONES

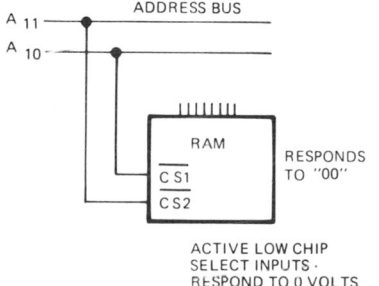

ACTIVE LOW CHIP
SELECT INPUTS ·
RESPOND TO 0 VOLTS,
THAT IS, ZEROES

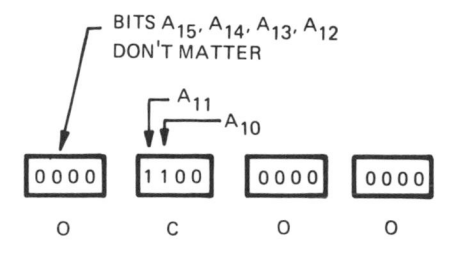

bus are both "1," that is, both at +5 volts. Some chip select inputs respond instead to 0 volts, and remain unactivated if they receive +5 volts; these are called "active low inputs," and they're designated by a bar over the pin designation, such as $\overline{CS1}$ or $\overline{CS2}$. Those which are normal, or "active high," and respond to +5 volts, are named without the bar, as for instance just "CS3."

All that's needed, then, is to connect bits A11 and A10 to normal, active high chip select inputs on the ROM chip, and the other ten lower address bits to the ROM's internal address inputs, and then make sure that, in addressing any other chips, we don't cause the address bits A11 down through A0 to take on values in the range that would activate the ROM. If we make sure that the combinations we use in bits A11 and A10 aren't the same for any other device, then we won't accidentally address the ROM.

Notice that bits A15, A14, A13, and A12 then don't matter. They don't need to be connected to any of the chips, because we have successfully used just A11

and A10 to differentiate the four
chips. That has a peculiar conse-
quence: since the first four bits
of the address don't matter (that
is, the first hex digit doesn't matter), the microprocessor can address one of the chips (say, the
ROM) with any of 16 different
first digits, and it will still work
the same way. For instance, we
have decided that the ROM will
run from 0C00 to 0FFF; but
since the first hex digit doesn't
matter, the microprocessor could
address it as 1C00 to 1FFF, or
2C00 to 2FFF, and so on. That
will turn out to be very convenient, because there will be a reason why it is good to have the
microprocessor   "think"   that
while the RAM is always in the
range 0000 to 0100, the ROM is
sometimes in the range 0C00 to
0FFF, and sometimes in the
range from FC00 to FFFF.

First let's see why the RAM
should start at 0000. The programs in the ROM are always re-
ferring to data in the RAM; so
the programs must include addresses for things in the RAM. It
would be an advantage to keep
those addresses as short as possible, to save ROM space. The
M6800 has some short-form in-

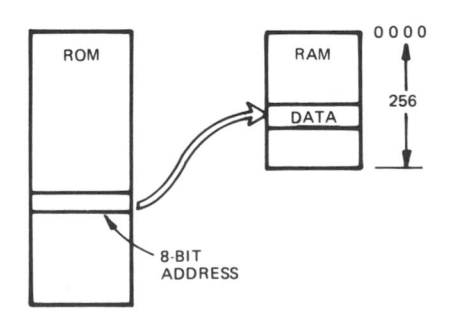

structions that save this program space, if you can arrange things so that the RAM data is in the first 256 locations of memory — that is, starting at 0000. Then, for those instructions, you need only 8 bits of address stored with the instruction rather than 16.

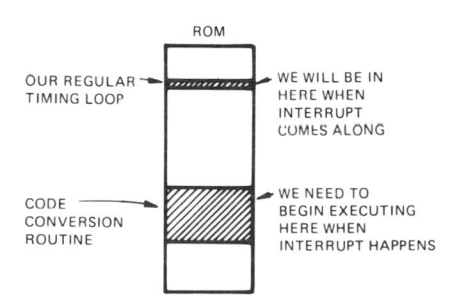

ROM

OUR REGULAR TIMING LOOP

WE WILL BE IN HERE WHEN INTERRUPT COMES ALONG

CODE CONVERSION ROUTINE

WE NEED TO BEGIN EXECUTING HERE WHEN INTERRUPT HAPPENS

Now let's see why the ROM "sometimes" should start at FC00. When the interrupt from the comparator comes along, we want the microprocessor to stop whatever it's doing — that is, stop timing the integrator — and begin immediately to convert the number in accumulator B to the proper codes to send out to the 7-segment displays. Now the instructions in the ROM to do those things will begin at some other place in the ROM, and the microprocessor has to be told where they are. In the M6800, this is done in a special way: when the interrupt signal comes along, the microprocessor always looks in locations FFF8 and FFF9 to get the 16 bits of address that tell it where to begin executing the interrupt routine, as the part of the program we want to execute now is called. So, we should store, in locations FFF8 and FFF9, the address of

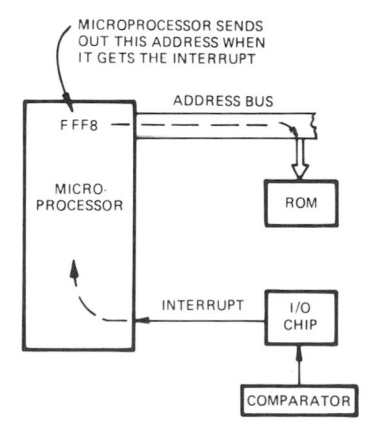

MICROPROCESSOR SENDS OUT THIS ADDRESS WHEN IT GETS THE INTERRUPT

ADDRESS BUS

FFF8

MICRO-PROCESSOR

ROM

INTERRUPT   I/O CHIP

COMPARATOR

the start of that part of the pro--
gram. And, of course, we want
to store that address right in the
ROM, along with the program.

But we said that the ROM goes
from 0C00 to 0FFF. How can
FFF8 and FFF9 be in that
range? It turns out that, because the first digit doesn't matter the way we have the ROM
wired, the microprocessor can
call for FFF8 and FFF9, and
what it will get is what is stored
in 0FF8 and 0FF9. So that's
where we put the address.

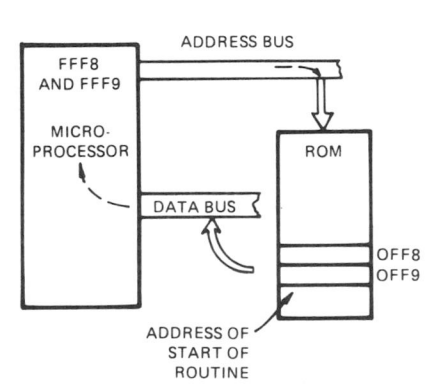

Naturally, the reason we get a-
way with this is that we don't
need very much of the address
space — we actually need only
1288 locations out of the total
possible 65,536. If we needed all
65,536 for real memory, we
couldn't do this.

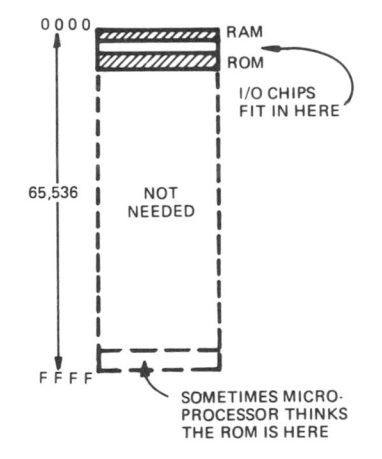

Of those 1288 locations, 1024
are for the ROM, 256 are for the
RAM, and the other 8 we will
use for the I/O chips.

Let's look, then, at how we
would actually wire up the ROM
and the RAM to the address bus.
As the diagrams show, we leave
the highest four bits of the address bus unconnected; bits A11

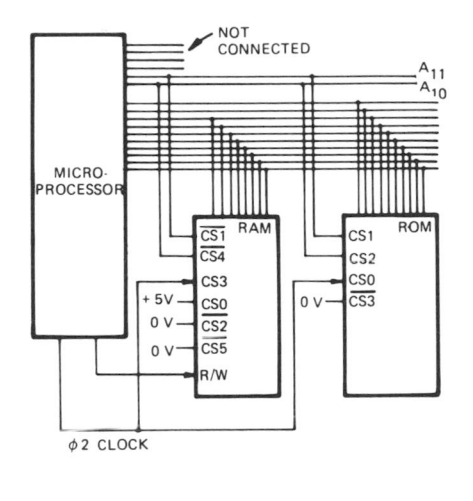

and A10 are connected to the various chip select inputs; and bits A9 through A0 are connected to the devices' internal address lines. Of the latter, the RAM needs only A7 through A0, because it has only 256 locations.

As we said before, we connect bits A11 and A10 to active high chip select inputs on the ROM, so that the ROM will be selected when those bits are both 1's, that is, at +5 volts.

However, we want to select the RAM when both those bits are 0 volts. So we connect A11 and A10 to two active low inputs on the RAM. Now it will respond to "00" on those lines.

But what happens when the microprocessor isn't trying to access the ROM or the RAM or the I/O chips? What happens, for instance, when the microprocessor is attempting to place an address on the address bus, but hasn't got it settled down yet?

To prevent the ROM and the RAM from being activated at the wrong time, we connect the phase 2 clock signal ($\phi2$) to active-high chip select inputs on

both chips. Now neither will be activated during the phase 1 clock — only when the phase 2 clock signal goes to +5 volts. And, at that time, the memory addresses will be good — they'll have settled down.

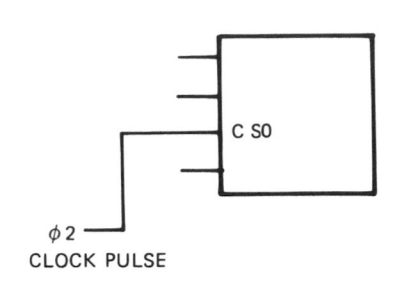

$\phi 2$
CLOCK PULSE

Where we have extra chip select inputs left over, we connect them to +5 volts if they're the active-high kind, and to 0 volts (ground) if they're the active-low kind. Then those chip select inputs will always be activated; and the chips will then react whenever the bits A11 and A10 are right for them AND the phase 2 clock says the addresses are good.

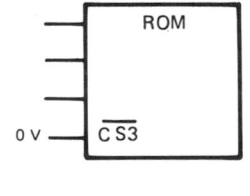

The RAM, however, in contrast to the ROM, can be both read and written into. But it will not "know" which we want to do just from the address bits alone. So there is provided, as an output from the microprocessor, a read/write signal (R/W) which we connect to the RAM. As we will see, it is also connected to the I/O chips, since in general they can be used for both input and output without rewiring them — though we are not using them that way.

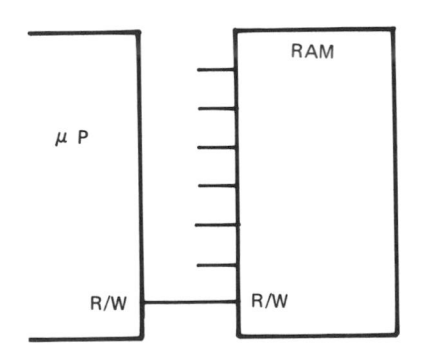

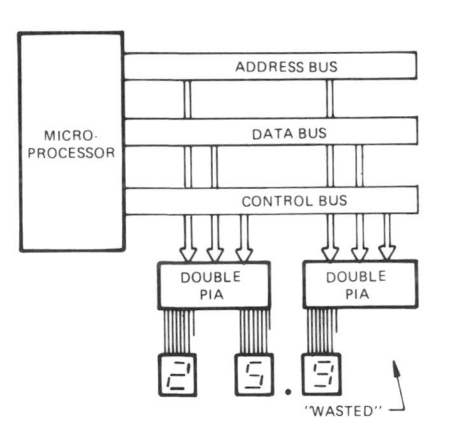

Now let's consider how the I/O chips are connected and addressed. In the block diagram we developed in chapters 4 and 5, we showed three I/O chips, one for each digit. There are, in fact, a number of different kinds of chips one can use for this purpose, but whichever one is chosen, it will have certain peculiarities that set it apart from other chips that could have been used.

The I/O chips that the Motorola Corporation offers for use with the MC 6800 are called "peripheral interface adapters," or PIAs. Each one is really a double I/O chip: it has two complete sets of eight lines each that you can connect to the "outside world." In our case, then, we need only one-and-a-half PIAs to service our three 7-segment displays (the other half will just have to be

"wasted," but, as is usually the case with microcircuits, that will be less expensive than buying three separate single I/O chips). Actually, we'll be able to use that other half to help control the integrator and comparator.

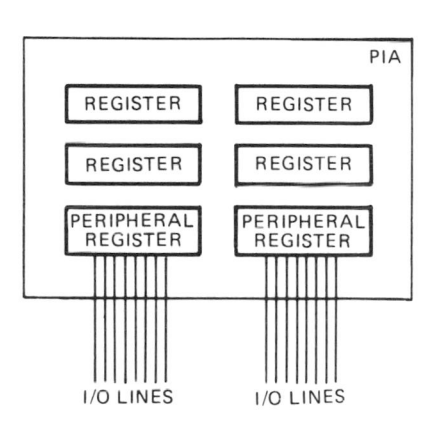

I/O LINES   I/O LINES

Because each PIA is double, it has two data holding registers, or "peripheral registers," as they are called, one for each set of input/ output lines. However, there are also two other registers associated with each peripheral register, which makes a total of six for each double PIA. Exactly what these registers are for, what data you need to send to each one, and how each one is addressed, are the questions we now need to discuss.

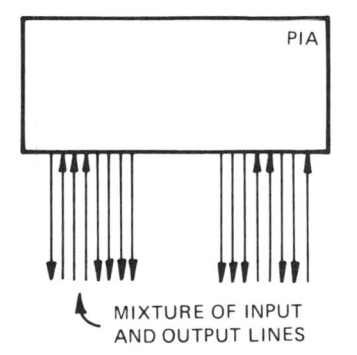

MIXTURE OF INPUT
AND OUTPUT LINES

The manufacturers of the PIA wanted to make it as universally usable as possible, because in that way the most chips could be sold, and the unit price could be kept low. They knew that some customers would want mostly input lines; others, mostly output lines, as in our application; others, a mixture of input and output lines; and still others, a mixture that could change — that is, a mixture in which any particular line could be an input line

some of the time, and an output
line at other times, all under the
control of the microprocessor.
So the PIA was designed with
this in mind.

To accomplish that, the designers
of the PIA added a "data direction register," the bits of which
control the directions of the in-
put/output lines. A data direction register has eight bits, one
for each input/output line, and
when you want a particular I/O
line to be an input line, you
make the corresponding bit of
the data direction register a "0."
A "1" in a bit of the data direction register makes the corresponding I/O line an output line.
All that's needed, then, is to de-
cide what pattern of input and
output lines you need at the moment, and send a group of eight
bits of that pattern to the data
direction register. By using the
word "10001111," for instance,
you can make five output lines
and three input lines, in the pattern given by the word.

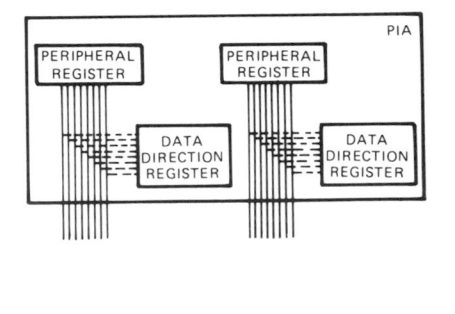

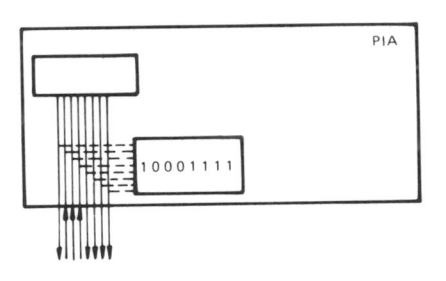

Again, to make the PIA chip as
useful as possible to as many different people as possible, the designers decided to build in some
interrupt control circuits, with

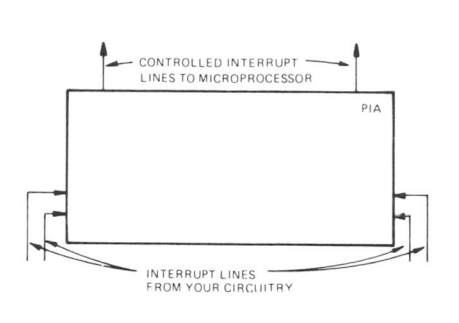

which the user can control up to four different interrupt lines, coming from various parts of his application. With the chip, he can enable or disable them individually, and control whether the interrupt happens when the external line goes high or goes low. We'll use one of these interrupt control circuits for our interrupt from the comparator.

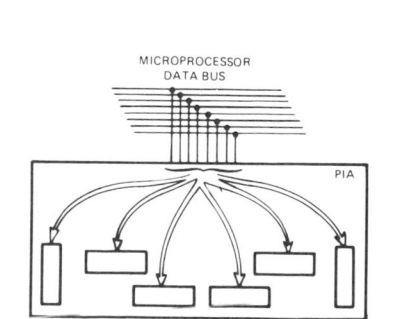

However, to allow the user some control over all that, the designers built in two control registers, each eight bits long; most of those bits are to control the interrupt circuits. The various options available to you, and how you have to set those bits to get them, are quite involved, and the usual method of dealing with them is to look them up in tables to find the option you want. When we discuss our interrupt, we'll show how we have selected the set of functions we needed.

Now let's ask: if we have to send data to six different registers inside the PIA, and there is only one eight-bit set of connections to the microprocessor's data bus, how do we select each of the six registers individually?

The PIA has two register select
lines devoted to this purpose. We
know, however, that two bits of
"address" can select only four
possibilities, and there are six
registers to be distinguished.
Another bit is needed; then we
would have a total of three bits,
which would give us eight possibilities, even more than we need.

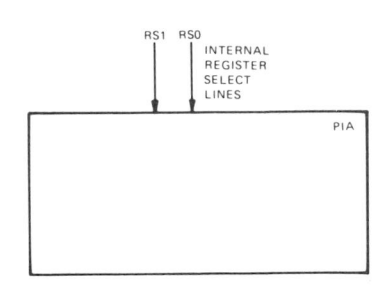

The PIA designers' solution to
this problem was influenced by
the fact that they just didn't
have an extra pin to spare; so
they put the third bit into the
control registers. They set it up
so that you can address one of
the control registers with just the
two external select lines ("01"
for control register A, "11" for
control register B); then, by sending in a control register word
that has a "1" or a "0" in the
third bit from the right (bit no.
2), you can direct data either to
one of the peripheral registers or
to one of the data direction registers.

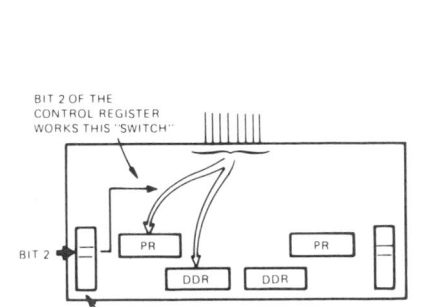

Complex? Yes. Quite normal,
however, in microelectronic circuits where internal complexity
is often used to try to save just
one external pin on the chip.

The two external lines that you use to select these registers are called RS1 and RS0. The "switch bits" in the control registers are called CRA-2 (bit number 2 of control register A) and CRB-2 (in control register B). The table of how these external lines and switch bits are set, in order to get at each of the six registers, is as follows, where "x" means that it doesn't matter which way that signal is set.

RS1	RS0	CRA-2	CRB-2	REGISTER SELECTED
0	0	1	x	PERIPHERAL REGISTER A
0	0	0	x	DATA DIRECTION REGISTER A
0	1	x	x	CONTROL REGISTER A
1	0	x	1	PERIPHERAL REGISTER B
1	0	x	0	DATA DIRECTION REGISTER B
1	1	x	x	CONTROL REGISTER B

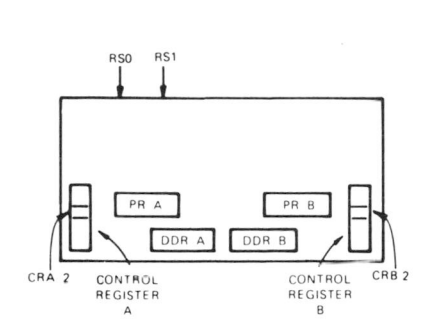

Now that we know how that works, let's see how we would connect the PIAs to the address bus, and what the addresses of the different registers inside the PIA would come out to be.

In addition to the internal register select lines RS1 and RS0, each PIA also has some standard chip select lines: CS0, CS1, and CS2, the latter being active low.

With these lines we can cause the whole PIA chip to be distinguished from other whole PIA chips, and from other chips on your address bus.

Remember that we intend to use bits A11 and A10 of the address bus to distinguish the four chips we have (one ROM, one RAM, and the two PIAs). Since we are using "00" for the RAM and "11" for the ROM, that leaves us "01" for the first PIA and "10" for the other. Remembering to connect to an active high chip select to recognize a "1," and to an active low chip select to recognize a "0," the correct connections are as shown.

Using "01" in bits A_{11} and A_{10} results in addresses that start at 0400 hex. Using "10" in bits A_{11} and A_{10} results in addresses that start at 0800 hex.

Now we can connect the select lines RS1 and RS0 to any of the lower order address lines, but for simplicity we choose to connect them to the two rightmost lines, A_1 and A_0. Running through the four combinations of these two bits, 00, 01, 10, and 11, causes the rightmost hex digit to be 0,

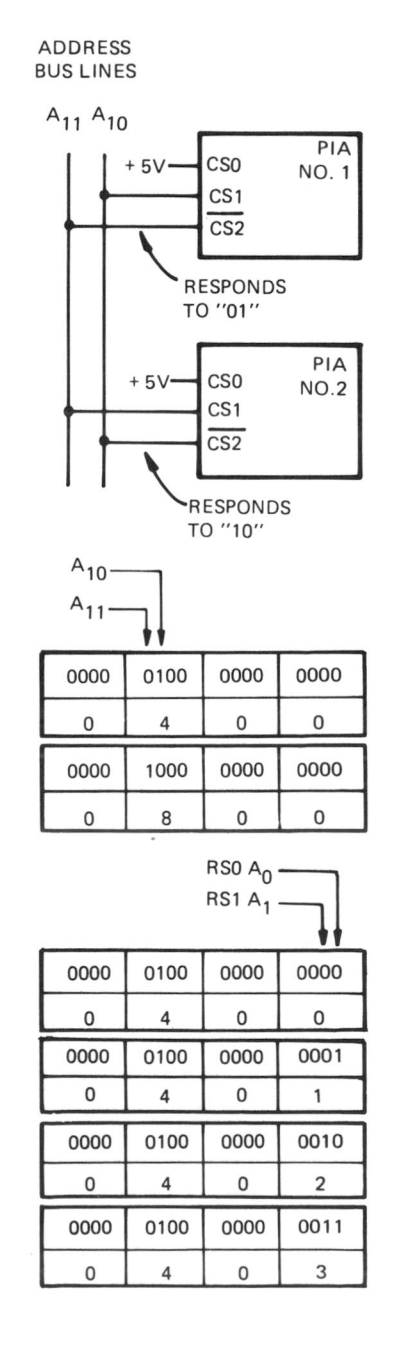

ADDRESS BUS LINES

A_{11} A_{10}

PIA NO. 1

RESPONDS TO "01"

PIA NO. 2

RESPONDS TO "10"

A_{10}
A_{11}

0000	0100	0000	0000
0	4	0	0

0000	1000	0000	0000
0	8	0	0

RS0 A_0
RS1 A_1

0000	0100	0000	0000
0	4	0	0
0000	0100	0000	0001
0	4	0	1
0000	0100	0000	0010
0	4	0	2
0000	0100	0000	0011
0	4	0	3

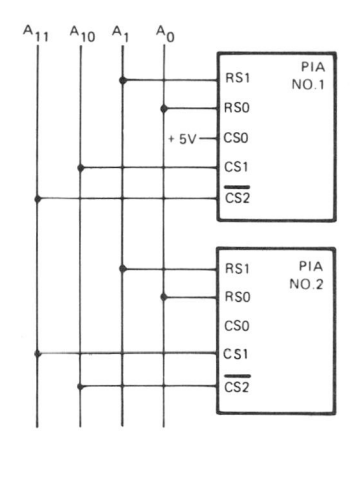

1, 2, and 3 respectively. Therefore: one PIA can be addressed
as 0400 through 0403 hex, and
the other as 0800 through 0803
hex.

It will help to visualize all this if
we assign special names for these
addresses:

PIA1AD 0400 PIA# 1, side A,
                       peripheral and
                       data direction
                       registers

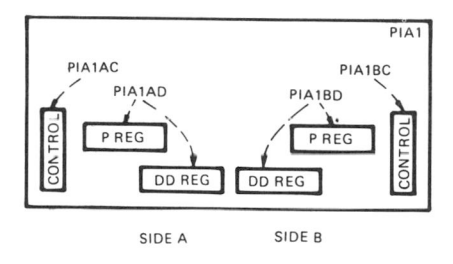

SIDE A   SIDE B

PIA1AC 0401 PIA # 1, side A,
                       control register

PIA1BD 0402 PIA # 1, side B,
                       peripheral and
                       data direction
                       registers

PIA1BC 0403 PIA # 1, side B,
                       control register

PIA2AD 0800 PIA # 2, side A,
                       peripheral and
                       data direction
                       registers

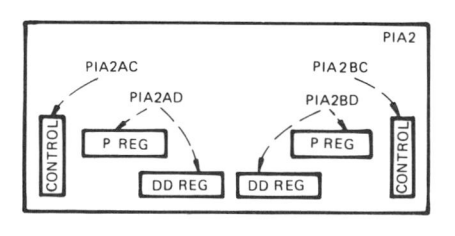

SIDE A   SIDE B

PIA2AC 0801 PIA # 2, side A,
                       control register

PIA2BD 0802 PIA # 2, side B,
                       peripheral and

data direction
registers

PIA2BC   0803 PIA # 2, side B,
control register

Notice that, after the letters PIA
in the address names, there is a 1
or a 2 indicating which PIA it is;
then a letter A or B indicating
which side of the PIA it is;
then a D or a C, telling whether
it is a peripheral/data direction
register or the control register.

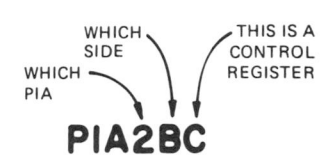

Our assembler program can accept address names — or labels,
as they are called — up to six
characters in length; so we can
use these labels in our program.
We "tell" the assembler program
how we want these labels to be
matched up with real addresses
in this way:

```
        ORG $0400
PIA1AD  RMB 1
PIA1AC  RMB 1
PIA1BD  RMB 1
PIA1BC  RMB 1
        ORG $0800
PIA2AD  RMB 1
PIA2AC  RMB 1
PIA2BD  RMB 1
PIA2BC  RMB 1
```

The "ORG" means "origin"

and the "$0400" means "hex 0400." This statement tells the assembler program to begin assigning addresses at 0400. This is followed with the four labels, each with the command (to the assembler program) to reserve memory space one byte long (RMB). The first one, PIA1AD, will get address 0400, as we want, and then the rest will be assigned succeeding addresses, again as we want. The same process works for the group starting at 0800.

```
        ORG $0400
PIA1AD  RMB  1
```

LOOK LIKE INSTRUCTIONS
BUT THEY AREN'T

Now these kinds of commands to the assembler program — they are called "assembler directives" — are not to be confused with instructions that our microprocessor will eventually execute. They are written right into your program, as if they were such instructions, but in reality they are just communication between you and the assembler. They do not result in any commands being placed in the ROM. But when you say "PIA1AD" elsewhere in your program, the assembler will now know enough to substitute the address "0400."

Now let's look at the instructions we have to have our microproces-

sor execute when we first turn it
on, to transmit the right information into the PIA control registers and data direction registers.

First we want to set all four data
direction registers so that all lines
are output lines; that is, we want
to put "1111 1111" in all four
direction registers.

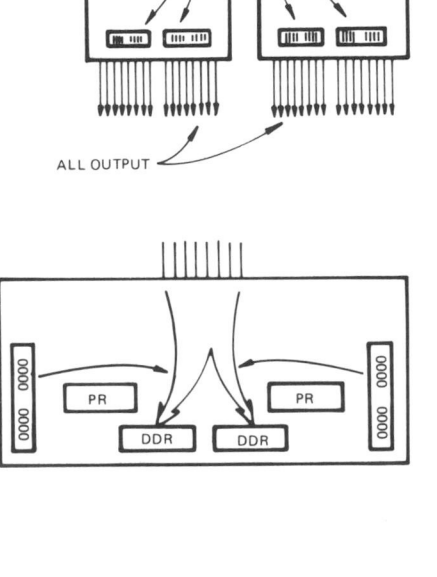

DATA DIRECTION
REGISTERS WITH
"1111 1111"

ALL OUTPUT

But to transmit data into the
data direction registers, we have
to be sure that the control registers all have zeroes in bit 2. So
first we would transmit, say,
"0000 0000" into the control registers, and then we can send
"1111 1111" into all the data
direction registers.

In the MC6800, sending information to such registers, now that
we have addresses for them, is a
simple matter. First you do a
"load accumulator" instruction,
putting the bit pattern you want
into one of the accumulators
(from the ROM). Then you do a
"store" instruction, mentioning
the label for the address of the
register into which you want the
data to go. That's all there is to it.

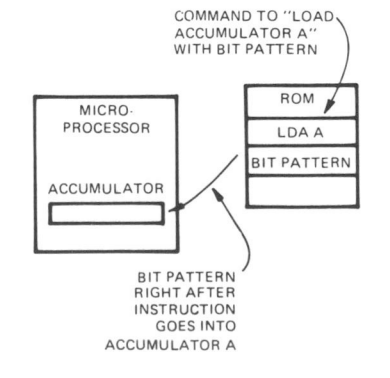

COMMAND TO "LOAD
ACCUMULATOR A"
WITH BIT PATTERN

MICRO-
PROCESSOR

ROM

LDA A

BIT PATTERN

ACCUMULATOR

BIT PATTERN
RIGHT AFTER
INSTRUCTION
GOES INTO
ACCUMULATOR A

To set the control registers properly, we first do:

LDA   A $00
STA   A   PIA1AC
STA   A   PIA1BC
STA   A   PIA2AC
STA   A   PIA2BC

Remember that #$00 means that we want hex 00, that is, binary 0000 0000, stored in the ROM right after the load (LDA A) instruction; so it will be picked up and loaded into the accumulator A. The four store instructions send the data 0000 0000 out to the four control registers.

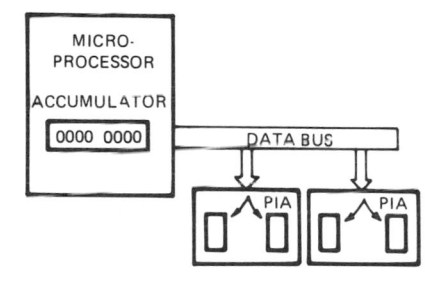

Now we can send the "1111 1111" to the four direction registers:

LDA   A   #$FF
STA   A   PIA1AD
STA   A   PIA1BD
STA   A   PIA2AD
STA   A   PIA2BD

Now that that's done, we want to reset our "switch bits" in the control registers, so that the codes for the 7-segment displays will go out to the peripheral registers (and so out to the displays) rather than to the data direction registers. To do that, we need to

set bit 2 of the control registers
(associated with the three displays) to a "1." Now bit 2 is the
third bit from the right, and to
make it a 1, we should send 0000
0100, which is hex "04." So we
send:

```
LDA  A   #$04
STA  A   PIA1AC
STA  A   PIA1BC
STA  A   PIA2AC
```

The last control register, PIA2BC,
we are going to use to control
our interrupt from the comparator; we will explain later how to
fill it with the right bits.

We are now all set up to send
data to the PIAs, and out to the
displays, whenever we want. All
we have to do is to load accumulator A with the data — in our
case, a 7-segment code we will
have retrieved from our ROM table — and then do a store instruction transferring the information
out to the appropriate peripheral
register in the PIAs.

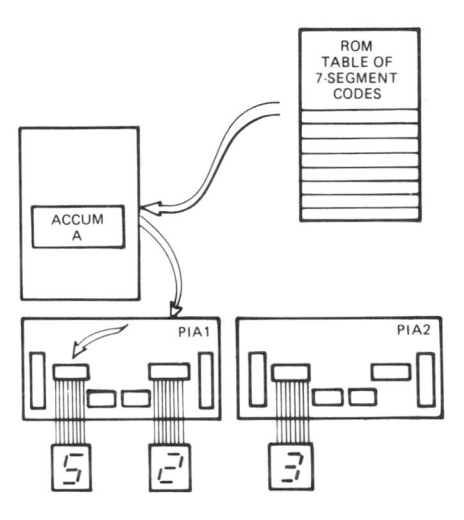

```
STA  A   PIA1AD
```

Getting the information out of
the ROM table is another story,
to which we will come presently.

11 CONTROLLING THE INTEGRATOR

ONE SIGNAL
TO RESET AND
START

ON/OFF

INTEGRATOR

v

t

RAMP

TO MICRO-
PROCESSOR

COMPARATOR

POT

Part of the problem in our application is the control of the integrator; we need to turn it on to
start the ramp, and we need to
turn it off when the comparator
tells us that its voltage has surpassed that of the potentiometer.
We can simplify that to some ex-
tent by building the integrator
circuit so that one signal, that is
either at +5 volts (for ON) or
0 volts (for OFF) does all the
controlling, and so that when the
integrator is turned on, it automatically starts out with its output (to the comparator) at zero
volts.

Then, one of the output lines
from one of the PIAs can do the
controlling. We've already used
both the A and B section of PIA
#1, and the A section of PIA #2,
to control the 7-segment displays;
that leaves the B section of PIA

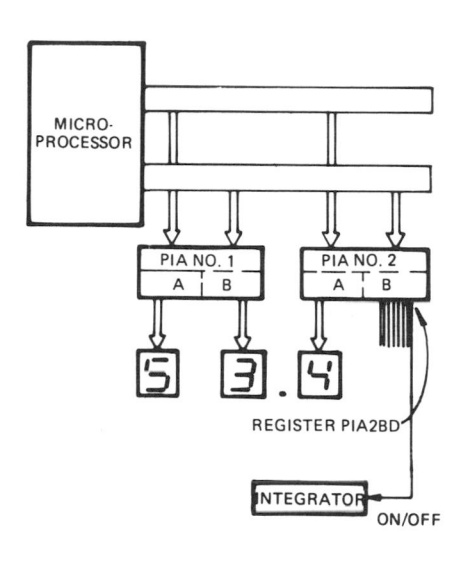

REGISTER PIA2BD

INTEGRATOR

ON/OFF

2. So we choose the rightmost bit of that section to control the integrator. To turn the integrator on, then, we need to send "0000 0001" to register PIA2BD.

That can be done with the following instruction sequence:

LDA A   #$01
STA A   PIA2BD

This loads accumulator A with hex 01 (binary 0000 0001) from the ROM; then transmits that data out to the desired register.

Later, after the signal comes into the microprocessor indicating that the comparator has "fired," we will want to turn the integrator off. To do that, first we clear accumulator A (which fills it with "0000 0000," and then send that data to PIA2BD:

CLR A
STA A   PIA2BD

The line to the integrator control will then revert to 0 volts, and the integrator will be turned off.

12

LOOKING UP
IN THE TABLES

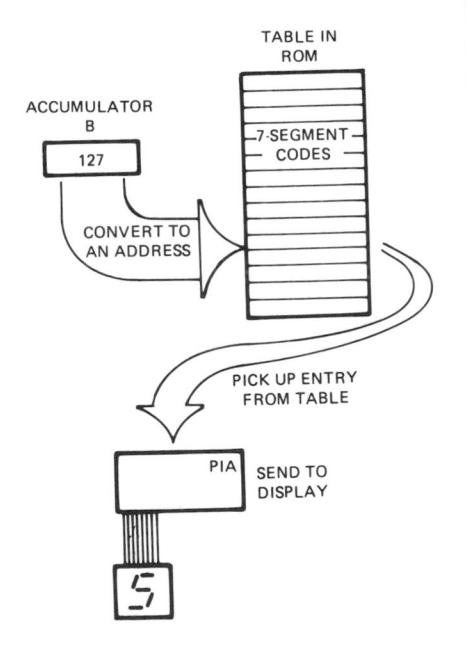

ACCUMULATOR B

127

TABLE IN ROM

7-SEGMENT CODES

CONVERT TO AN ADDRESS

PICK UP ENTRY FROM TABLE

PIA   SEND TO DISPLAY

So far, we've considered how to time out the integrator and to count that time with a number in accumulator B; and we've considered how to send the 7-segment codes to the displays. Between those two operations comes the other important function of our apparatus: the conversion of the number in accumulator B to the proper set of 7-segment codes, through a lookup process in a table in the ROM.

In any table lookup process in a computer, the strategy is to convert the number with which you're starting — the number in accumulator B — to an address, or set of addresses, at which the correct entry in the table can be found. In our case, that is complicated somewhat by the need to retrieve three separate 7-segment codes — one for the tens

digit, one for the units digit, and one for the tenths digit (here always "0" or "5").

If we only had to retrieve one code, we could do that by arranging the codes in the table in ascending sequence, and by starting the table out at some address that had "0000 0000" as the last eight bits of the address itself. Then the 7-segment code that corresponded to the number "0" in accumulator B would be at the address that ended in "0000 0000"; the code that corresponded to "1" in accumulator B would be at the address that ended in "0000 0001"; and so on. We need to take care of the numbers 0 through 199 in accumulator B; so the address of the last one would end in "1100 0111," the binary equivalent of 199.

Addresses are, however, 16 bits long, and we need to have a whole address to get at the codes. So we would just take the starting address of the table — let's say we picked "0C00" — and add to it the number in accumulator B. The result would be the correct full address of the code we needed. If the number

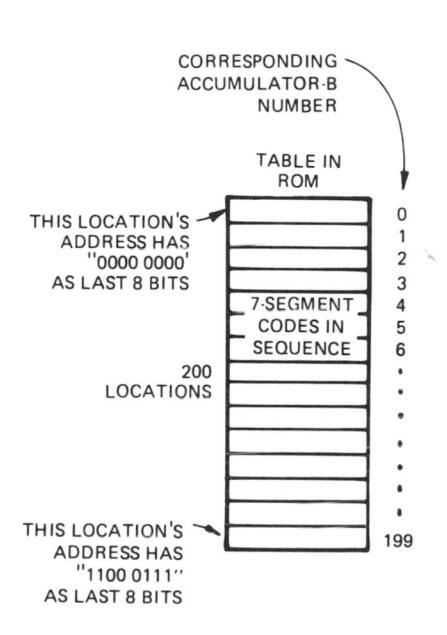

CORRESPONDING ACCUMULATOR-B NUMBER

TABLE IN ROM

THIS LOCATION'S ADDRESS HAS "0000 0000" AS LAST 8 BITS

7-SEGMENT CODES IN SEQUENCE

200 LOCATIONS

THIS LOCATION'S ADDRESS HAS "1100 0111" AS LAST 8 BITS

0
1
2
3
4
5
6
•
•
•
•
•
•
•
199

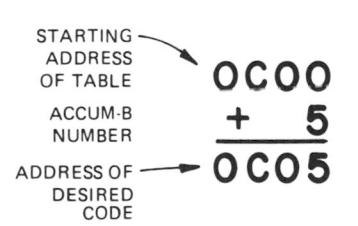

STARTING ADDRESS OF TABLE

ACCUM-B NUMBER

ADDRESS OF DESIRED CODE

0C00
+   5
0C05

in accumulator B were zero, the address would come out "0C00"; if the number were "5," the address would come out "0C05." If the number were "199," the address would come out (in full binary) 0000 1100 1100 0111, that is, 0CC7 hex. Notice that the last eight bits are what we said they ought to be.

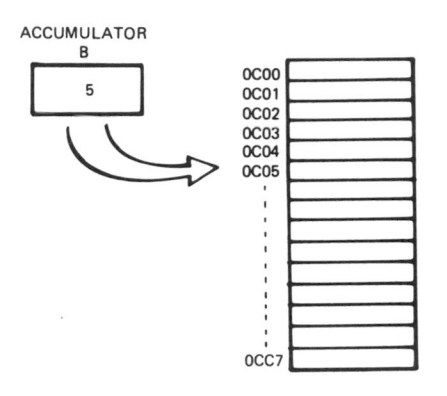

We said that we had to <u>add</u> the number in accumulator B to the starting address of the table. But if the number in accumulator B never goes over 255 (in our application it only goes to 199) — and if the table starts out at a location the last eight bits of which are 0000 0000 (as is the case in address 0C00), then we can do this adding just by taking the first eight bits of the starting address, and the eight bits of the accumulator B number, and just pushing them together. The "0C" ("0000 1100") will be on the left end, and the bits from accumulator B will be on the right end. By carefully choosing the right kind of starting address and limiting our table to less than 256 locations, this trick has become possible. And, as we will see, it will save us some program space in our microcomputer, which does not have the capa-

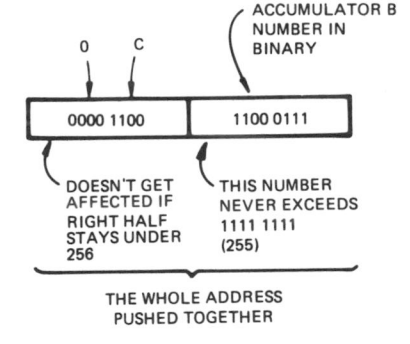

bility to add two 16-bit numbers directly (it would have to accomplish that by a rather lengthy ROM-consuming subroutine).

But all of that assumed that we had only one code to look up; and we really have three codes to look up for a certain number in accumulator B. How can we do that and use the same trick?

The strategy we choose to use is to have three separate tables: one for the tens digit, one for the units digit, and one for the tenths of gallons digit. We will start each table out at an address the last eight bits of which are "0000 0000." The tens digit table will start at 0C00; the units digit table at 0D00; and the tenths table at 0E00. Now the proper three addresses at which to pick up our three codes can be found by "pushing together" 0C with the contents of accumulator B, 0D with those contents, and 0E with those contents. Then a "load accumulator" command, using one of those addresses, will place one of the 7-segment codes into the accumulator A, and from there a store operation is all that is needed to send it out to the PIA and on to the display.

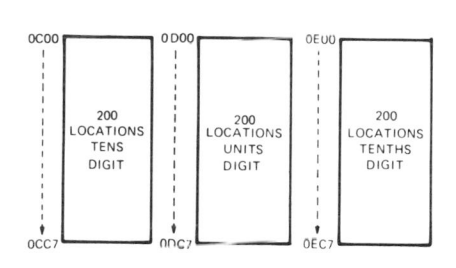

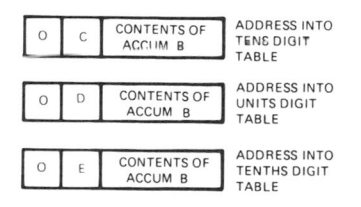

Now, how can we "push together" those two sets of bits? To
do that, we will make use of a
property of the index register in
the MC6800. The instruction
"LDX," followed by an address,
causes to the microprocessor to
go to that address, get its contents and put it into the left half
of the index register; then to go
to the next location after that,
get its contents, and put them into the right half of the index register. So if we could put the
"0C," for instance, into some
location (say location 0000 in
the RAM), and then put the contents of accumulator B into location 0001 in the RAM, and
then carry out this "LDX" command using 0000 as the address,
we would get both halves of our
address loaded into the index
register.

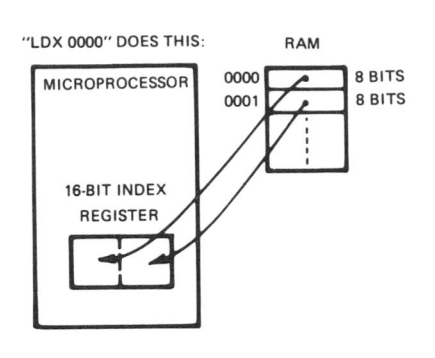

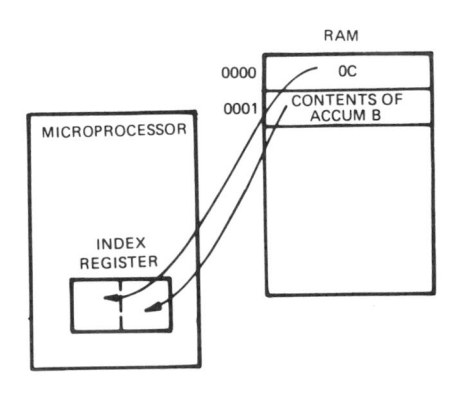

Finally, there is a form of the
load-accumulator instruction in
which you can ask for the contents of the index register to be
used as the address. You simply
write, in assembly language,

LDA   A   X

which means, "load accumulator
A with whatever is at the address
given in the index register."

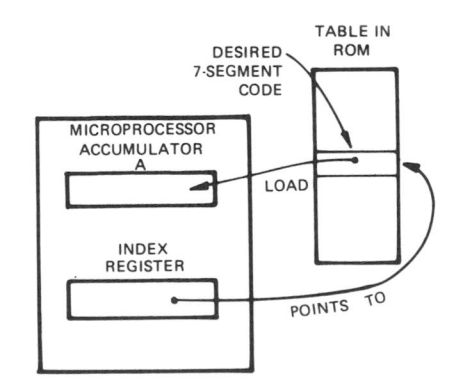

In our application, what is at the location given in the index register is the 7-segment code we want.

Exactly the same operation can be carried out for "0D" and "0E," and we can get all three 7-segment codes.

Now let's show exactly how we would do it. First, we have to tell the assembler program where the three tables are:

```
ORG    $0C00
RMB    200
ORG    $0D00
RMB    200
ORG    $0E00
RMB    200
```

Next, we set up the three "staging areas" in the RAM where we will push the addresses together:

```
       ORG    $0000
TABH   RMB    2
TABM   RMB    2
TABL   RMB    2
```

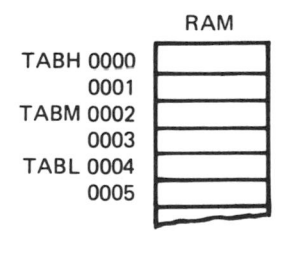

RAM

TABH 0000
0001
TABM 0002
0003
TABL 0004
0005

This sets up a pair of adjacent locations 0000 and 0001, and calls 0000 "TABH," a label we have invented meaning "table high"; then two more locations at 0002 and 0003, with 0002 called

"TABM" (table middle") and two more at 0004 and 0005, with 0004 called "TABL" ("table low").

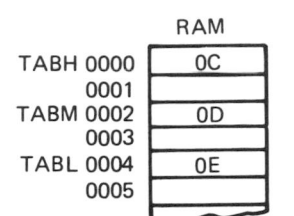

Next, we have to fill in the "left halves" of our addresses: the "0C," "0D," and the "0E." Now this must be done after the microcomputer begins to execute, because these staging areas we have set up are in RAM — which always loses its data when the machine is turned off. So we put, in the stream of instructions that the microcomputer must execute just after it is turned on,

LDA A   #$0C
STA A   TABH
LDA A   #$0D
STA A   TABM
LDA A   #$0E
STA A   TABL

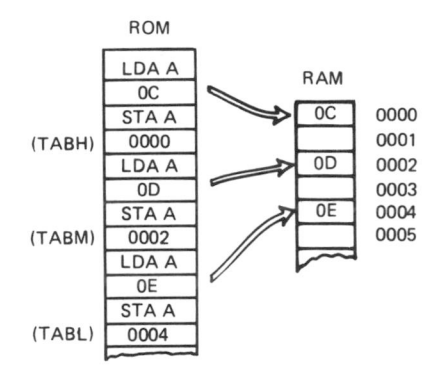

What this does is to cause the microcomputer to pick up the "0C," "0D," and "0E" from ROM locations (embedded right in the stream of instructions), and transfer them to the RAM locations TABH, TABM, and TABL (that is, 0000, 0002, and 0004). That only has to happen once, right after the microcomputer is turned on; after that, so

long as power is not lost, that data will be in the RAM at those locations.

All that remains is to put the right halves of the addresses in place. That, however, occurs in a different part of the program: the part that is executed after the integrator's voltage has just exceeded that of the potentiometer, and that therefore the number in accumulator B is the one we need to convert to codes. We will treat that part of the program next.

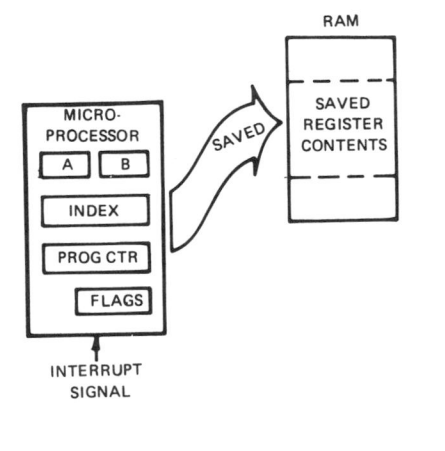

The basic idea of an interrupt is simple: an external device, like our comparator, generates a voltage on a signal line, which causes the microprocessor to suspend operations (at the conclusion of the current instruction), save the contents of all its internal registers (by putting them into RAM), read a prearranged location in ROM to find out where it should begin executing, and then begin to execute at that place.

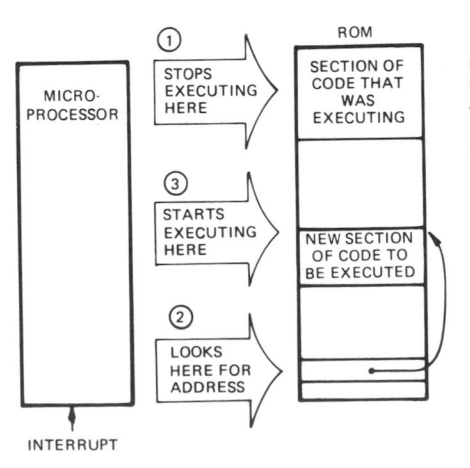

In our application, "that place" will be the start of a program section in which the table lookup of the 7-segment codes is done, and in which those codes are transmitted out to the displays. After that, the microprocessor will transfer back to the "main program," in which the integrator is reset and turned on to climb the ramp again.

How is the comparator actually connected to the interrupt input of the microprocessor? It's done using the one of the PIA chips as an intermediate step, because then the interrupt signal can be controlled so that it happens when we want it to, and not inadvertently while we are trying to set up at the beginning of the execution.

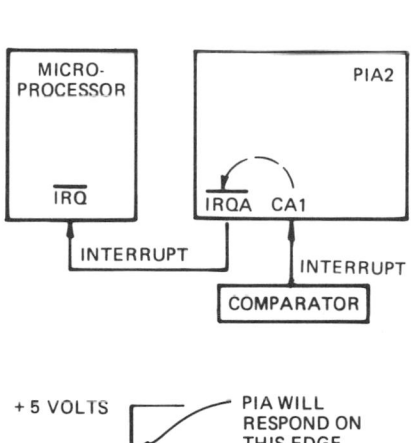

The signal line from the comparator is actually wired to a pin called "CA1" on PIA #2. Then the $\overline{IRQA}$ output of PIA #2 is wired to the "$\overline{IRQ}$" input of the microprocessor: the letters "IRQ" mean "interrupt request," and the bar over the letters means that the signal is active low (0 volts means ON).

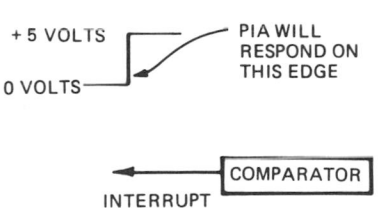

Now we can control when the $\overline{IRQ}$ signal is sent to the microprocessor by the PIA: we want it sent when the signal from the comparator goes from 0 volts to +5 volts. We control that by making the two rightmost bits of the control register on the A side of PIA # 2 equal to "11." Other combinations of those two bits can disable the $\overline{IRQ}$ signal entirely, or make it respond when CA1 goes from high to low.

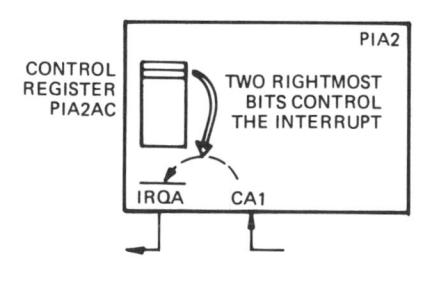

So we should send "0000 0011" to the control register, PIA2AC, when we want to enable the interrupt (allow it to come through). However, as you may remember, we were already using the third bit from the right in this register to control the passage of data into the peripheral register rather than the data direction register; for that we needed that bit to be "1." Consequently what we really need to send is not "0000 0011" but "0000 0111."

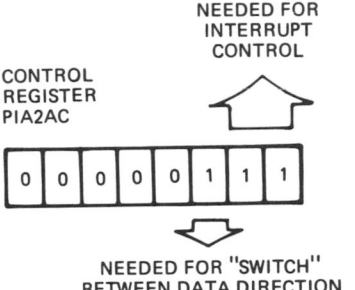

NEEDED FOR
INTERRUPT
CONTROL

CONTROL
REGISTER
PIA2AC

NEEDED FOR "SWITCH"
BETWEEN DATA DIRECTION
REGISTER & PERIPHERAL REGISTER

That is hex 07; so we would execute these instructions:

```
LDA  A    #$07
STA  A    PIA2AC
```

— and that's all there is to it.

When the interrupt comes in, then, the microprocessor will respond, as we will describe in detail.

14

RESPONDING TO THE COMPARATOR'S INTERRUPT

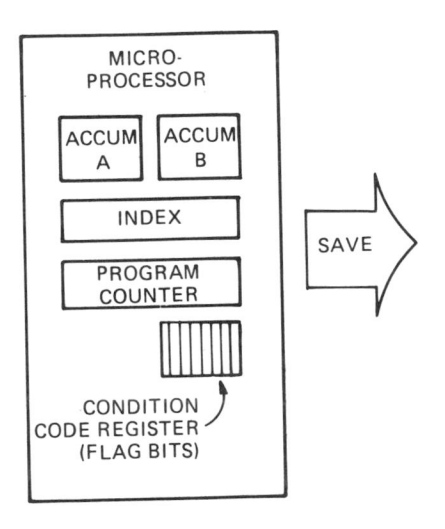

When the comparator's signal comes on, it enters pin CA1 on PIA #2, and is passed on via signal $\overline{IRQA}$ to the microprocessor. As soon as the microprocessor has finished whatever instruction it happens to be executing, it will begin its interrupt sequence, which starts with some actions to save the contents of the internal registers: accumulator A, accumulator B, the index register, the program counter, and the condition code register (the register that contains all the flag bits).

As we indicated before, the contents of these registers are saved in the RAM. Later, when a special "return from interrupt" command is executed, they will all be retrieved from RAM and inserted back in their proper registers. That way, the microprocessor can take up exactly where it left off (if that is the best thing to do).

Now this saving is done in a special part of the RAM called the "stack." It's special only in that we have designated that part of the RAM for use as the stack. Writing into the stack, and reading back from it, however, are controlled in a special way; the addresses used are found in the "stack pointer register," another special register in the microprocessor.

The usual way to explain a stack is by an analogy to a spring-loaded dish storage device like those used in restaurants: you put the dishes in one at a time, and their weight makes the stack of dishes sink into the well, so that the last one in is the first one in position to come back out. That's exactly the way a computer stack works: as a word is written into the stack in the RAM, using the stack pointer register's contents as the address, the stack pointer register is decremented, so that the next word will go into the next lower location ("above" it in the memory .) So the stack pointer keeps track of the next location into which a word should go in the stack.

All this is entirely automatic in the MC6800; those registers like the program counter, that have 16 bits, are stored away in two locations automatically, and when the process is reversed, the right data is found and put back into those registers.

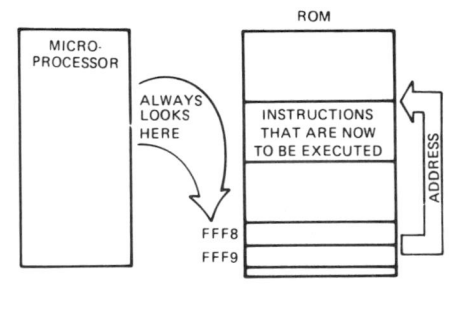

After all these registers are saved, the microprocessor then needs to know where to begin executing in order to "service" the interrupt. It always "looks" in the same place to find this address: it looks in location FFF8 (and the succeeding location, FFF9). We must, therefore, put the address of the starting location — of the part of the program we want executed after the interrupt — into locations FFF8 and FFF9.

Remember, though, that we have deliberately wired the address bus connections so that the first hex digit of the address does not matter. If the microprocessor puts "FFF8" out on the address bus, our ROM will respond just as if that were "0FF8," which is in the range of our ROM. So we can put this starting address in 0FF8 and 0FF9, and everything will work.

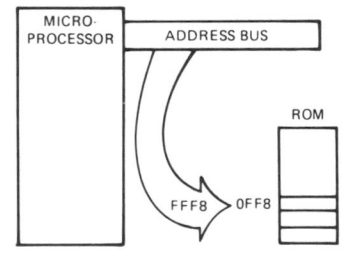

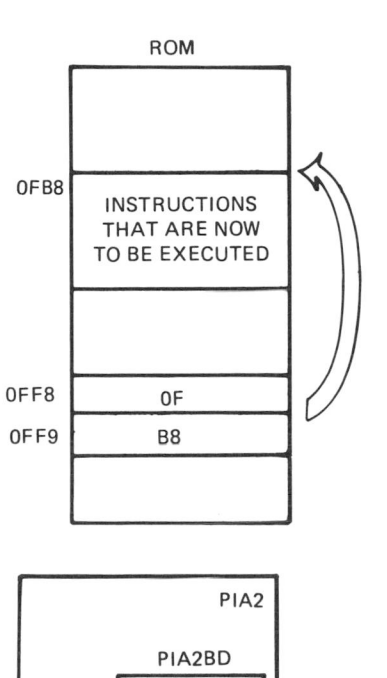

ROM

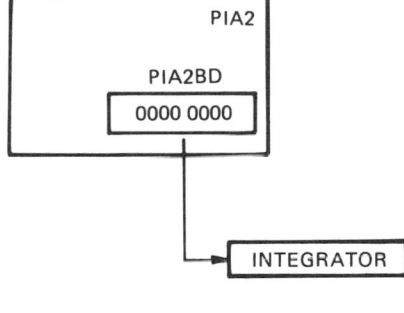

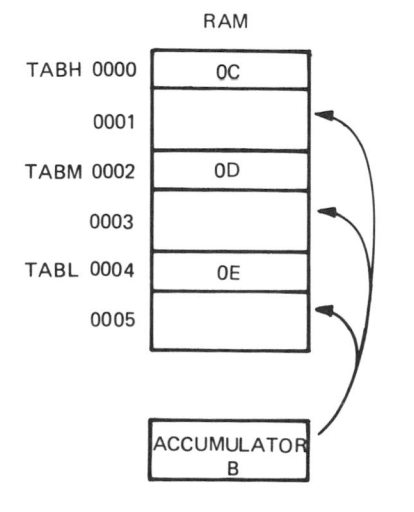

We choose to have this part of the program start at 0FB8; so we would tell the assembler program to put "0F" in 0FF8, and "B8" in 0FF9, this way:

ORG $0FF8
FCB $0F, $B8

The assembler directive "FCB" tells the assembler program to fill in "0F" at the first location specified by the ORG statement, and "B8" in the next location.

What, now, should the instructions at location 0FB8 be — the instructions that begin converting our number to display digits?

First, they should shut off the integrator:

CLR   A
STA   A   PIA2BD

This sends "0000 0000" to the peripheral register of PIA #2 side B, the one in which the rightmost bit controls the integrator.

Next, the contents of accumulator B must be "pushed together" in the RAM with the starting addresses of the three lookup tables. That is done by placing those

contents into the locations immediately following TABH, TABM, and TABL:

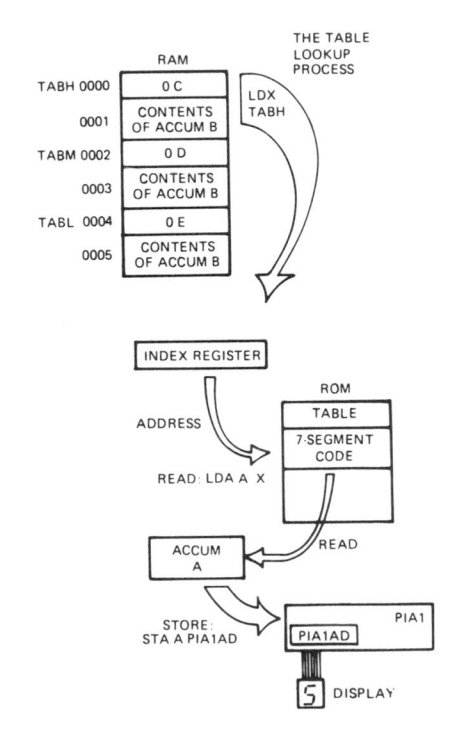

    STA B TABH + 1
    STA B TABM + 1
    STA B TABL + 1

That forms up the three addresses in which the high, medium, and low digits of our tank reading lie. Now the tables must be read using those addresses. As we described before, that's done by first transferring the addresses into the index register with an LDX instruction, then loading accumulator A using the index register as the ROM address (which actually does the retrieving from the ROM table), and finally storing what's in accumulator A (the 7-segment code) out through the PIA to the display:

    LDX   TABH
    LDA A   X
    STA A   PIA1AD

    LDX   TABM
    LDA A   X
    STA A   PIA1BD

    LDX   TABL
    LDA A   X
    STA A   PIA2AD

That completes the table lookup operation and the transmitting of

the 7-segment codes to the displays. All that is left now is to transfer the microprocessor back so that it will start the integrator up again and begin counting up accumulator B for the next cycle.

Now there is an instruction in the MC6800, "RTI," or "return from interrupt," which will retrieve all the register contents from the stack, and begin executing where it left off. "Where" is determined, of course, by the program counter, which always contains the address of the next instruction to be executed. However, the program counter is one of the registers that was saved in the stack, and the contents that were saved give the next address after the last instruction executed before the interrupt. That's not really where we want to begin again.

Where we really want to begin is at location 0F30, the start of the integrator routine. So before coming out of our interrupt routine, we will "fool" the microprocessor by changing the contents of the stack, making the place in the stack where the program counter was stored read "0F30."

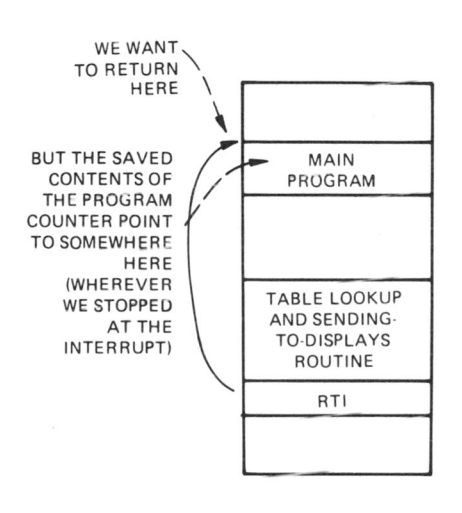

WE WANT TO RETURN HERE

BUT THE SAVED CONTENTS OF THE PROGRAM COUNTER POINT TO SOMEWHERE HERE (WHEREVER WE STOPPED AT THE INTERRUPT)

MAIN PROGRAM

TABLE LOOKUP AND SENDING-TO-DISPLAYS ROUTINE

RTI

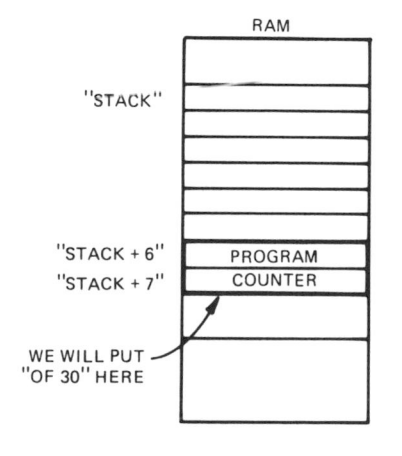

RAM

"STACK"

"STACK + 6"   PROGRAM
"STACK + 7"   COUNTER

WE WILL PUT "0F30" HERE

We do that this way:

LDA A   #$0F
STA A   STACK + 6
LDA A   #$30
STA A   STACK + 7

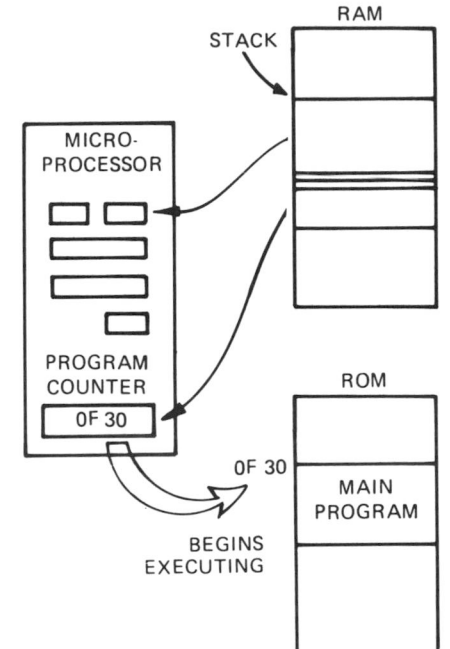

That puts "0F30" in the correct part of the stack so that when we do the RTI instruction,

RTI

and the stack is retrieved from RAM, the program counter will begin at 0F30, right where we want it to.

15

SOME "HOUSEKEEPING"

We've shown you what the major electrical connections and the major computer instructions need to be, for our small application. But, as in every task of this sort, there are some auxiliary connections and instructions that have to be included, to service the peculiarities of the equipment we've chosen to do the job.

When addresses are being put out onto the address bus, they are not "stable" for a brief period (typically 150 nanoseconds, or about one-seventh of a microsecond, in the M6800, for instance). It's important that the various chips tied to the address bus should not react at this time, since the addresses are changing erratically and could trigger the wrong chip. In the case of the ROM and the RAM, the phase 2 clock ($\phi2$) signal can be used to

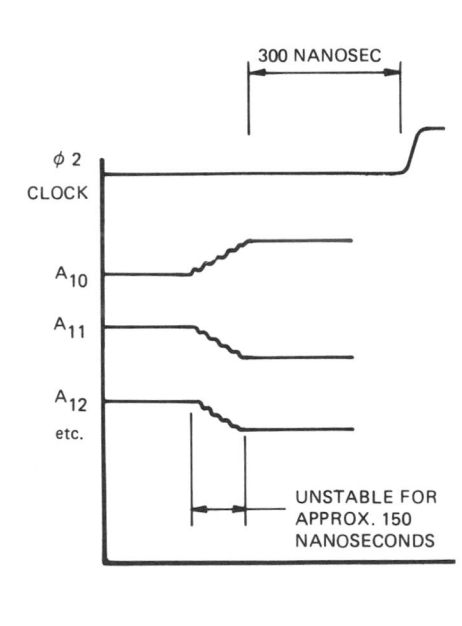

300 NANOSEC

$\phi2$
CLOCK

A_{10}

A_{11}

A_{12}
etc.

UNSTABLE FOR APPROX. 150 NANOSECONDS

solve the problem. It is OFF, or down at zero volts, during this interval, and comes back up to +5 volts about 300 nanoseconds after the addresses have completely stabilized.

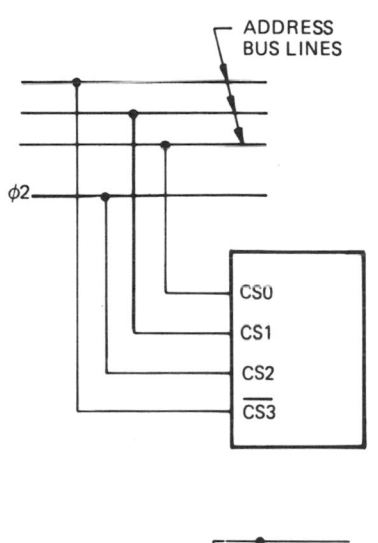

By connecting the $\phi 2$ clock to the ROM chip and the RAM chip — that is, by connecting it to those chips on chip select lines that require a +5v level to be activated, the chips can be prevented from reacting until $\phi 2$ comes on, after the rest of the address is stable.

There are a few conditions in the MC6800 that can cause an address to go out on the address lines even when $\phi 2$ is low. It turns out that this will not harm the operation of the ROM or the RAM, but those conditions could cause the PIAs to miss an interrupt signal, like the one coming in from the comparator. So another special signal, VMA ("valid memory address") is provided, which can be connected to PIAs to prevent this from happening.

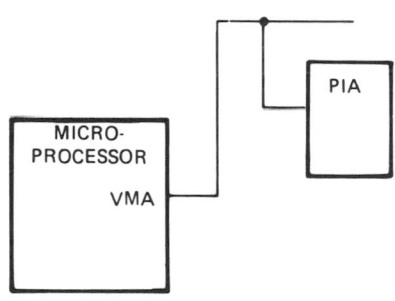

As we indicated before, the RAM needs a signal to inform it whether we are trying to write into it or simply read it. The

signal which does this is the read/write, or R/W signal, coming out of the microprocessor; it is controlled by the instructions to read or to write (load and store, that is, LDA and STA). This must be connected to the R/W input of the RAM.

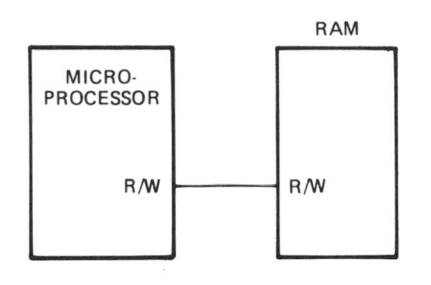

That takes care of the other electrical connections needed for "housekeeping." There are, as we said, some instructions of the housekeeping type which have to go into our program.

First of all, at the beginning of the actual running instructions of our program, we want to have some instructions that set everything up, or "initialize" our small system. We've already mentioned the transferring of "0C," "0D," and "0E" from ROM to RAM to act as the left-hand parts of the table addresses (we called them TABH, TABM, and TABL). And we talked about setting the PIA's so they would send data outward rather than inward, and then setting bit 2 of their control registers so we could send them data. In order to get these tasks done, we want the microprocessor to work away at them without the possibility of its being

CONDITION CODE OR
"FLAG" REGISTER

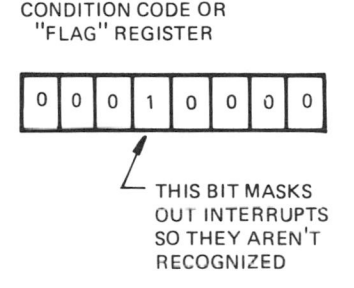

THIS BIT MASKS
OUT INTERRUPTS
SO THEY AREN'T
RECOGNIZED

falsely interrupted by the comparator. So first we use an instruction that makes the microprocessor ignore interrupts: we use the "TAP" instruction, to load up the condition code register (the register with all the flag bits in it) with the hex value "10," which is binary "0001 0000." That fifth bit from the right is the one that "masks" out the interrupt. Later, after we're finished with the setup phase, we can send binary "0000 0000" to that register, which will "enable" the interrupt (allow it to come through when it happens).

The instructions are, to mask out the interrupt,

        LDA   A   #$10
        TAP

and to "unmask" or restore its operation,

        CLR   A
        TAP

We also have to set the control register of PIA #2, side A, to allow the interrupt to come through, and to define which polarity of the incoming signal we want for the interrupt, as we covered in Chapter 13.

Then, we have to get the "stack" ready so that when the interrupt comes in, the data in the microprocessor's internal registers can be saved properly until after the interrupt routine is over. First, we will have defined where the stack is in the RAM:

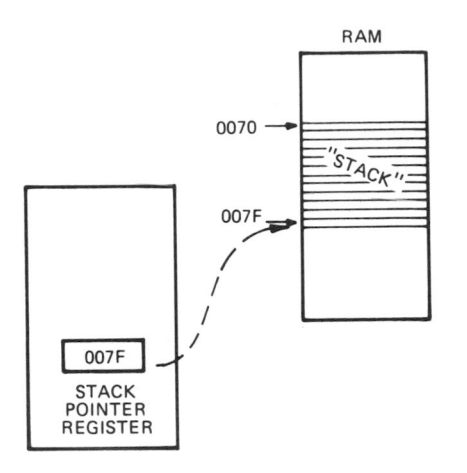

```
        ORG   $0700
STACK   RMB   12
```

Those, of course, are assembler directives, not executable statements. But in the setup phase, we want this executable instruction:

```
LDS   #$007F
```

This causes hex 007F to be placed in the stack pointer register. Notice that this address is the "bottom" of the stack, 15 locations down (that is, higher in addresses) than the "top" of the stack at 0070. When the interrupt happens, the microprocessor will begin transferring the contents of its internal registers to the bottom of the stack, and work its way upward (but going downward in memory addresses).

The order in which it does this is such that the contents of the pro-

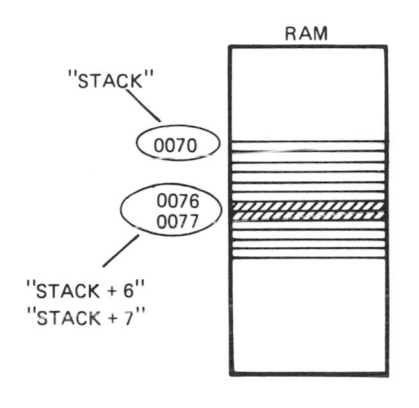

"STACK"

0070

0076
0077

"STACK + 6"
"STACK + 7"

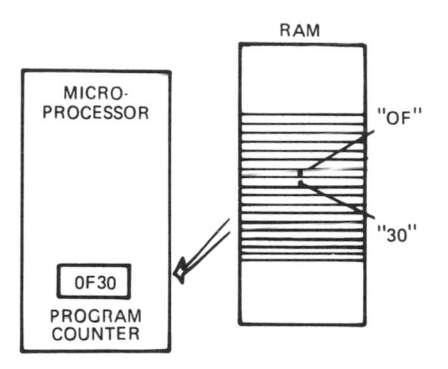

gram counter will wind up in 0076 and 0077, which, according to the assembler directives we gave above, would be STACK +6 and STACK +7. Down in the interrupt routine itself — the one in which the actual table lookup and the transmittal of the digits to the 7-segment displays takes place — we want to go in and modify this address. That's because, when we come back out of the interrupt routine, we definitely want to start at location 0F30, which is the proper start point for the routine that counts the integrator time. So, near the end of the interrupt routine, we execute:

```
LDA   A   #$0F
STA   A   STACK +6
LDA   A   #$30
STA   A   STACK +7
```

which properly modifies the location in the stack that will go back into the program counter. When it does, control will transfer to 0F30, and the microcomputer will take off from there.

Finally, we have to have some way for the microcomputer to begin at the right place when power is first turned on. The micropro-

cessor looks in locations FFFE and FFFF to find the starting address at which to begin executing. In our case, that should be the beginning of the setup phase, which is at 0F00. So we include in our data declarations section:

```
ORG   $0FFE
FCB   $0F,$00
```

which places "0F00" in locations 0FFE and 0FFF. Remember, though, that in our application, the first hex digit doesn't matter; so when the microprocessor looks in FFFE and FFFF, it will find the "0F00."

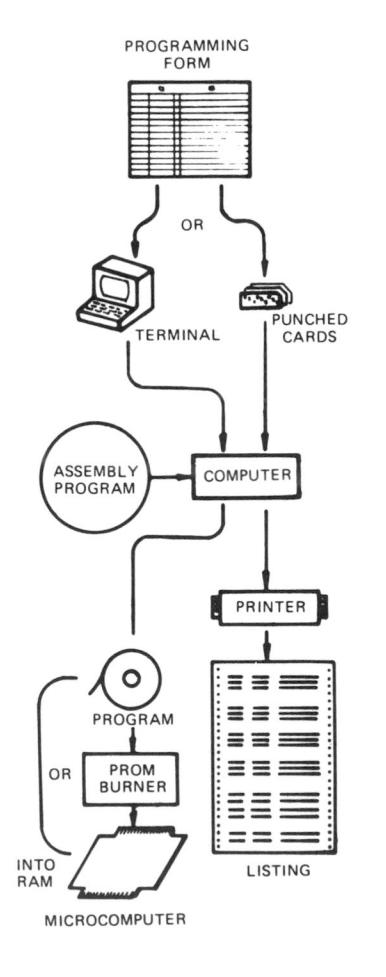

PROGRAMMING
FORM

OR

TERMINAL

PUNCHED
CARDS

ASSEMBLY
PROGRAM

COMPUTER

PRINTER

PROGRAM

OR

PROM
BURNER

INTO
RAM

LISTING

MICROCOMPUTER

That takes care of the "house-keeping" instructions, and now you can see the entire system put together. The block diagram, expanded to show all the pin connections, is given on the following page, followed by the full version of the Warnier-Orr diagram. Then comes the complete listing of the program itself, as printed out by the assembler program on a computer printer. The listing is divided into columns; they are as follows: first, a statement number, such as 10.000, which is assigned by the assembler program; then the address into which the instruction will go, over which you have control, but which is kept track of by the assembler program; then the "code," or actual instructions, in hex form (this is what will go right into the ROM or RAM); and finally, the rest of

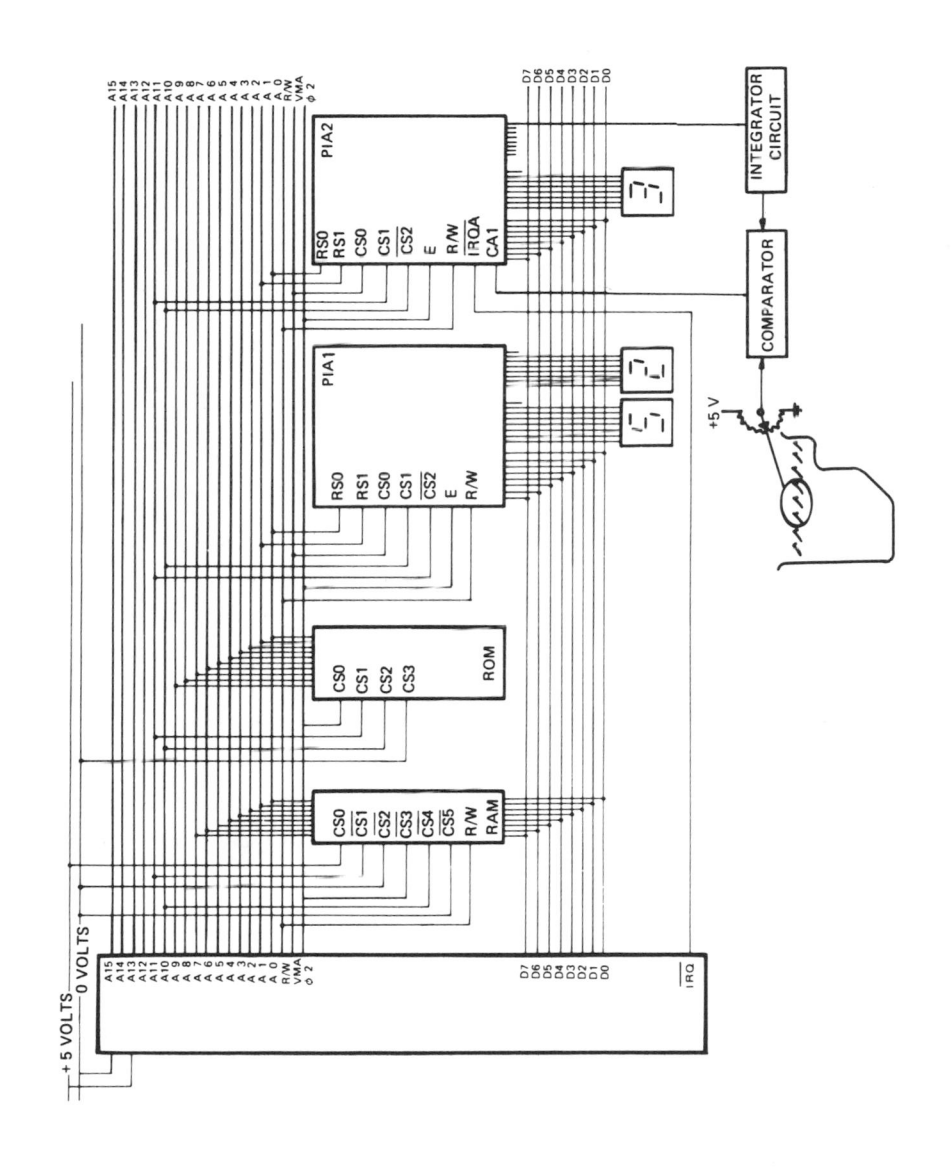

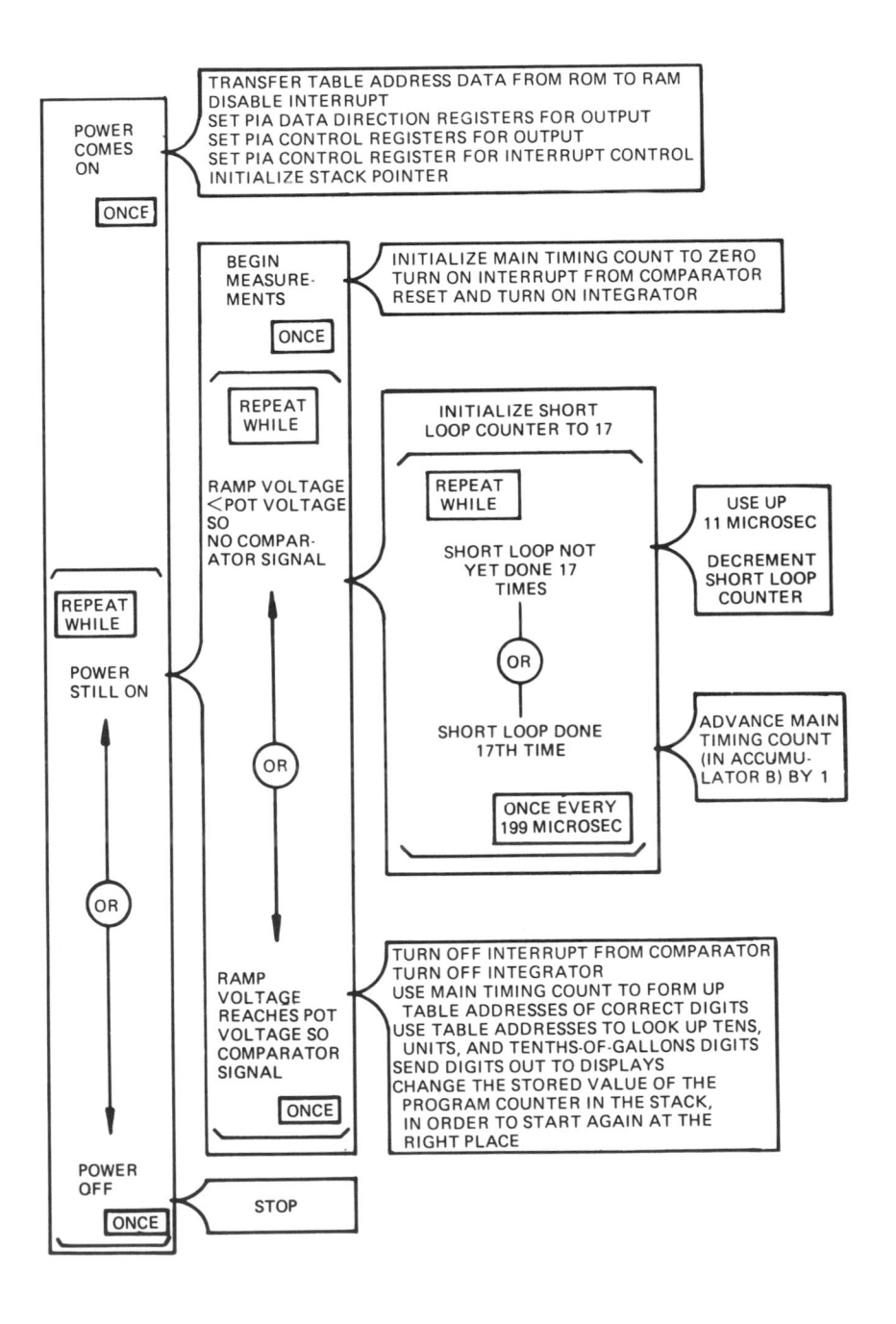

ADDRESSES INTO
WHICH INSTRUCTIONS
WILL GO

STATEMENT NO.
(ASSIGNED BY
ASSEMBLER)

ACTUAL
MACHINE
CODES IN
HEX

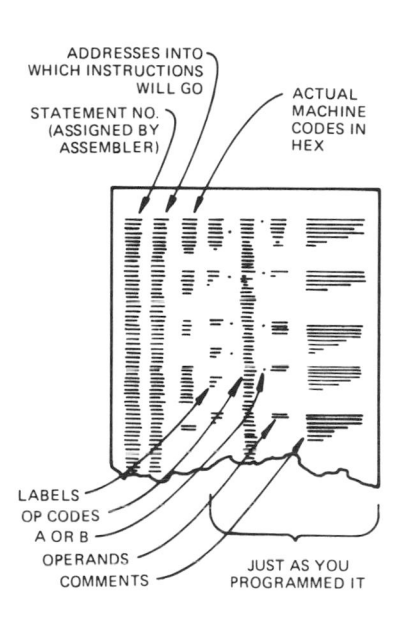

LABELS
OP CODES
A OR B
OPERANDS
COMMENTS

JUST AS YOU
PROGRAMMED IT

the columns are the program just as you wrote it on the keypunch forms, or typed it into a computer terminal. These columns contain the statement label (like "TIME" and "STEP"); the operation code ("LDX," "DEX," and so forth); the accumulator designation, A or B, where needed; and the operand. The operand can be actual data to be included with the instruction, or can be a label that is to be branched or jumped to, or can be the name you gave to a memory location that has the data or is to receive the data. Finally, there is a comments column.

In the comments column, we've added the number of the chapter in which you can find the explanation of the instruction.

You'll notice that the listing has been presented here in three pages, though it came from the computer printer as one long sequence. The first page contains the "declarations," in which we "told" the assembler program the names of all the relevant memory locations, where we wanted them in memory, and how much memory each one should take. The second page

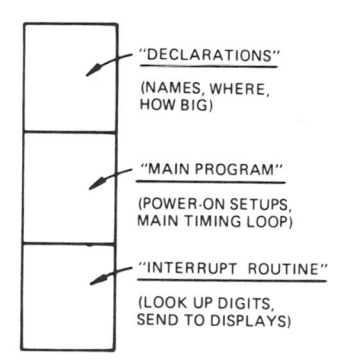

"DECLARATIONS"

(NAMES, WHERE,
HOW BIG)

"MAIN PROGRAM"

(POWER-ON SETUPS,
MAIN TIMING LOOP)

"INTERRUPT ROUTINE"

(LOOK UP DIGITS,
SEND TO DISPLAYS)

contains the "main" part of the program, that has in it the power-on setup operations, which transfer data from ROM to RAM where it's needed, send control information to the PIAs, and control the interrupt signals from the comparator and the start and stop signals out to the integrator; and then the "main timing loop," on which we spent such time in Chapter 7. Finally, the third page contains the "interrupt routine," that is, the part of the program that is done after the interrupt from the comparator comes in, shutting off the integrator, forming up the table addresses with the count in accumulator B, looking up the digits in the table, and finally sending the digits out to the displays.

STMT	ADDR	CODE	STATEMENT	
10.000			* M6800 PROGRAM IN ASSEMBLY LANGUAGE FOR OPERATION	
20.000			* OF THE LEVEL MEASURING SYSTEM	
30.000	0C00		ORG $0C00 ROM SPACE FOR 7 SEGMENT DISPLAY TABLE	⎫
40.000	0C00		RMB 200	
50.000	0D00		ORG $0D00	⎬ CHAPTER 12
60.000	0D00		RMB 200	
70.000	0E00		ORG $0E00	
80.000	0E00		RMB 200	⎭
90.000	0000		ORG $0000 RAM SPACE FOR TABLE LOOKUP ADDRESSES	⎫
100.000	0000		TABH RMB 2	
110.000	0002		TABM RMB 2	⎬ CHAPTER 12
120.000	0004		TABL RMB 2	⎭
130.000	0070		ORG $0070 STACK SPACE IN RAM	⎫ CHAPTER
140.000	0070		STACK RMB 12	⎭ 12
150.000	0400		ORG $0400 FIA REGISTERS	⎫
160.000	0400		FIA1AD RMB 1	
170.000	0401		FIA1AC RMB 1	
180.000	0402		FIA1BD RMB 1	
190.000	0403		FIA1BC RMB 1	
200.000	0800		ORG $0800	⎬ CHAPTER 10
210.000	0800		FIA2AD RMB 1	
220.000	0801		FIA2AC RMB 1	
230.000	0802		FIA2BD RMB 1	
240.000	0803		FIA2BC RMB 1	⎭
250.000	0FFE		ORG $0FFE RESTART ADDRESSES	⎫ CHAPTER
260.000	0FFE	0F00	FCB $0F,$00	⎬ 15
270.000	0FF8		ORG $0FF8	⎫ CHAPTER
280.000	0FF8	0FB8	FCB $0F,$B8	⎭ 13

STMT	ADDR	CODE	STATEMENT [CONT.]		
290.000			ORG	$0F00	FILL IN THE LEFT HALF OF TABLE ADDRESSES IN RAM — CHAPTER 12
300.000	0F00	86 0C	START LDA A	$0C	
310.000	0F02	97 00	STA A	TABH	
320.000	0F04	86 0D	LDA A	$0D	
330.000	0F06	97 02	STA A	TABM	
340.000	0F08	86 0E	LDA A	$0E	
350.000	0F0A	97 04	STA A	TABL	
360.000	0F0C	86 10	LDA A	$10	
370.000	0F0E	06	TAP		TURN OFF INTEGRATOR — CHAPTER 15
371.000	0F0F	86 00	LDA A	$00	SET CONTROL REGISTERS TO SEND DATA TO DATA DIRECTION REGISTERS — CHAPTER 10
372.000	0F11	B7 0401	STA A	PIA1AC	
373.000	0F14	B7 0403	STA A	PIA1BC	
374.000	0F17	B7 0801	STA A	PIA2AC	
375.000	0F1A	B7 0803	STA A	PIA2BC	
380.000	0F1D	86 FF	LDA A	$FF	SET PIA DATA REGS FOR OUTPUT — CHAPTER 10
390.000	0F1F	B7 0400	STA A	PIA1AD	
400.000	0F22	B7 0402	STA A	PIA1BD	
410.000	0F25	B7 0800	STA A	PIA2AD	
420.000	0F28	B7 0802	STA A	PIA2BD	
430.000	0F2B	86 04	LDA A	$04	SET BIT 2 IN CONTROL REG — CHAPTER 10
440.000	0F2D	B7 0401	STA A	PIA1AC	
450.000	0F30	B7 0403	STA A	PIA1BC	
460.000	0F33	B7 0803	STA A	PIA2BC	
470.000	0F36	86 07	LDA A	$07	SET CONT REG FOR INTERRUPT CONTROL — CHAPTER 13
480.000	0F38	B7 0801	STA A	PIA2AC	
490.000	0F3B	8E 007F	LDS	$007F	INITIALIZE STACK POINTER — CHAPTER 14
500.000	0F3E	5F	COUNT CLR B		TURN ON INTERRUPT — CHAPTER 15
510.000	0F3F	4F	CLR A		
520.000	0F40	06	TAP		
530.000	0F41	86 01	LDA A	$01	SET BIT TO RESET AND TURN ON INTEGRATOR — CHAPTER 11
540.000	0F43	B7 0802	STA A	PIA2BD	
550.000	0F46	CE 0011	TIME LDX	$17	INITIALIZE TIMING LOOP — CHAPTER 7
552.000	0F49	27 04	TEST BEQ	STEP	BRANCH IF LOOP COUNT = 0
553.000	0F4B	09	DEX		DECREMENT LOOP COUNT
554.000	0F4C	7E 0F49	JMP	TEST	REPEAT LOOP
555.000	0F4F	5C	STEP INC B		INCREMENT MAIN TIMING LOOP
556.000	0F50	7E 0F46	JMP	TIME	REPEAT TIMING ROUTINE

```
STMT       ADDR  CODE      STATEMENT [CONT.]

600.000    0FB8            ORG     $0FB8     START OF INTERRUPT ROUTINE                     } CHAPTERS
610.000    0FB8  4F        CLR A             SEND ZEROES TO PIA                             } 11, 13
620.000    0FB9  B7 0802   STA A   PIA2BD    SHUT OFF INTEGRATOR                            }
630.000    0FBC  D7 01     STA B   TABH+1    FORM TENS-DIGIT TABLE ADDRESS                  } CHAPTER
640.000    0FBE  D7 03     STA B   TABM+1    FORM UNITS-DIGIT TABLE ADDRESS                 } 13
650.000    0FC0  D7 05     STA B   TABL+1    FORM TENTHS-DIGIT TABLE ADDRESS                }
660.000    0FC2  DE 00     LDX     TABH      PUT TENS-DIGIT ADDRESS IN INDEX REGISTER       }
670.000    0FC4  A6 00     LDA A   X         GET TENS-DIGIT FROM TABLE                      }
680.000    0FC6  B7 0400   STA A   PIA1AD    SEND TENS-DIGIT TO PIA AND OUTPUT              }
690.000    0FC9  DE 02     LDX     TABM      PUT UNITS-DIGIT ADDRESS IN INDEX REGISTER      } CHAPTER
700.000    0FCB  A6 00     LDA A   X         GET UNITS-DIGIT FROM TABLE                     } 13
710.000    0FCD  B7 0402   STA A   PIA1BD    SEND UNITS-DIGIT TO PIA AND OUTPUT             }
720.000    0FD0  DE 04     LDX     TABL      PUT TENTHS-DIGIT ADDRESS IN INDEX REGISTER     }
730.000    0FD2  A6 00     LDA A   X         GET TENTHS-DIGIT FROM TABLE                    }
740.000    0FD4  B7 0800   STA A   PIA2AD    SEND TENTHS-DIGIT TO PIA AND OUTPUT            }
750.000    0FD7  86 0F     LDA A   #$0F      CHANGE STORED VALUE OF PROGRAM COUNTER         }
760.000    0FD9  97 76     STA A   STACK+6   IN STACK SO MICROPROCESSOR WILL RETURN         } CHAPTERS
770.000    0FDB  86 3E     LDA A   #$3E      TO 'COUNT' AND BEGIN NEW INTEGRATOR CYCLE      } 13, 14
780.000    0FDD  97 77     STA A   STACK+7                                                  }
790.000    0FDF  3B        RTI               RETURN TO 'MAIN PROGRAM'
```

ANOTHER WAY TO READ OUT

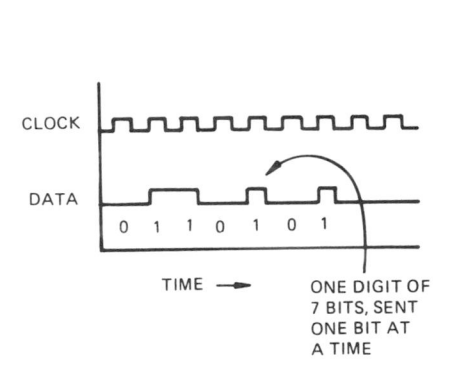

ONE DIGIT OF
7 BITS, SENT
ONE BIT AT
A TIME

Suppose that we had been asked to feed our three digits, representing the amount of liquid in the tank, to a printing device instead of to three 7-segment displays. Suppose further that, as is usual with small printing devices, this one accepts the data over a single wire (and a ground wire), in bit-serial format: first you send the seven bits representing the first digit, one bit at a time, in synchronism with a special clock signal that the printer provides. Then you send the second digit in the same way, and so on.

Microprocessor manufacturers commonly provide special I/O chips to help do that task, since it is a common one. In the case of the MC6800, the chip is called an Asynchronous Communications Interface Adapter, or ACIA, because one of its most

important uses is to change data from eight-bit bytes in parallel, to serial bit streams (one bit at a time on one wire), eventually to be sent over telephone or radio connections.

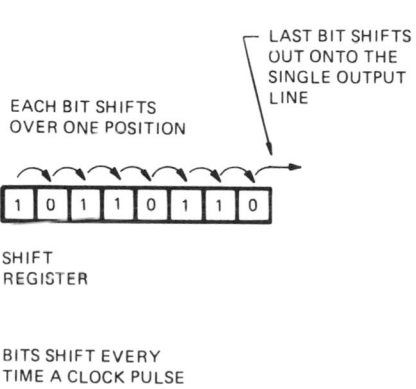

EACH BIT SHIFTS
OVER ONE POSITION

LAST BIT SHIFTS
OUT ONTO THE
SINGLE OUTPUT
LINE

SHIFT
REGISTER

The ACIA basically works by placing the eight-bit byte into what is called a shift register. In this device, the bits can be moved sideways one bit at a time each time the register is "shift-ed." The shifting is, of course, done in synchronism with a clock, which can be one which comes from the destination at the far end (as in our example of the small printer), or one which is generated locally.

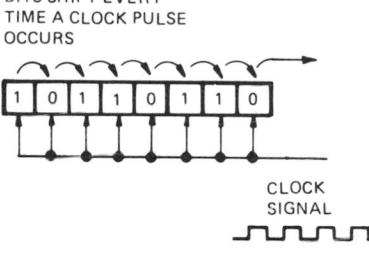

BITS SHIFT EVERY
TIME A CLOCK PULSE
OCCURS

CLOCK
SIGNAL

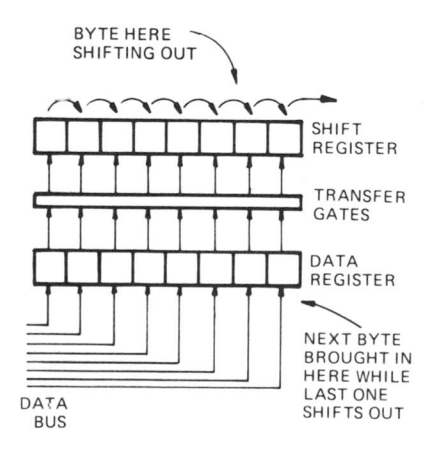

BYTE HERE
SHIFTING OUT

SHIFT
REGISTER

TRANSFER
GATES

DATA
REGISTER

NEXT BYTE
BROUGHT IN
HERE WHILE
LAST ONE
SHIFTS OUT

DATA
BUS

In addition to the shift register, the ACIA also has a data register which accepts the eight-bit byte from the microprocessor's data bus, and stores it until the shift register is ready to accept it. That tends to smooth out the transmission rate of the bits; while the shift register is shifting, and therefore tied up transmit-ting out to the destination, the data register can be "negotiat-ing" with the microprocessor for the next eight bits. When the shift register finishes, if the next

eight bits are already in the data
register, the movement of the
data from the data register to the
shift register is done automatically by the ACIA chip in time to
keep the bits flowing in an unin-
terrupted fashion.

As in the case of the PIA chip,
the ACIA has numerous registers,
all sharing the data bus route to
and from the microprocessor.
To save pins going into and out
of the ACIA, various commands
are given to the chip by transmit-
ting control bits into a register
inside the ACIA.

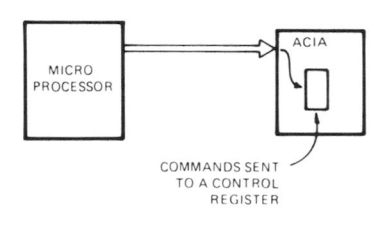

Since the ACIA can both receive
and transmit, and does so on
separate wires, it is provided with
two shift registers (each with its
associated data register), one for
receiving and one for sending.
There are, then, a Transmit Data
Register, a Transmit Shift Register, a Receive Data Register, and
a Receive Shift Register. In addition, there is a Control Register,
into which commands to the
ACIA are put, and a Status Register, into which the ACIA puts
information about the status of
various transmissions — which
can then be read by the microprocessor so it will know what to

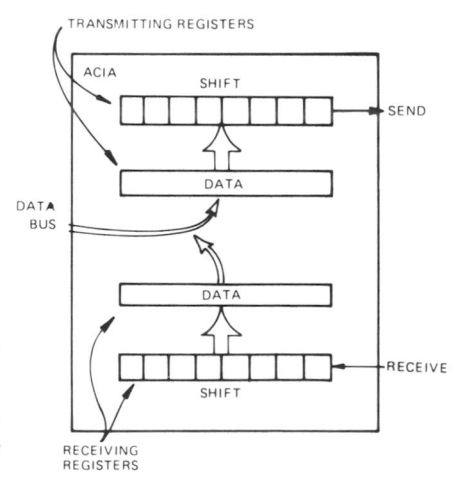

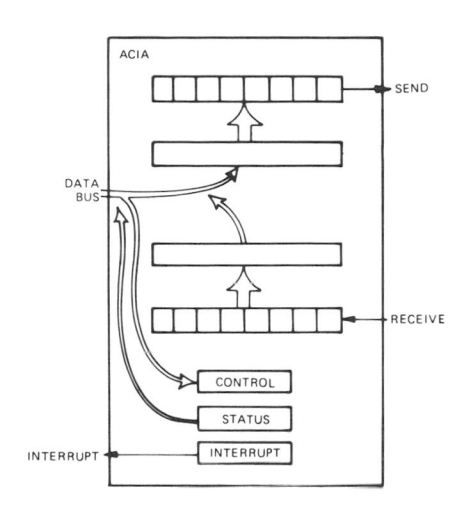

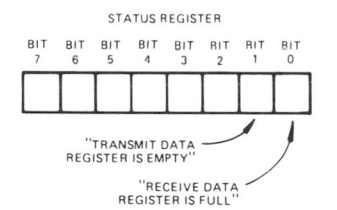

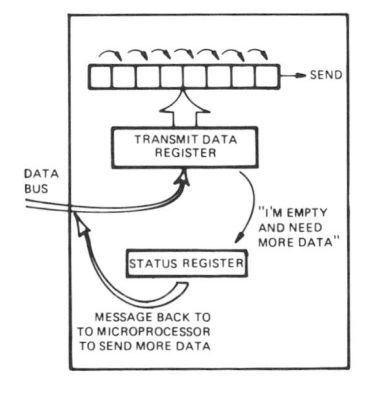

do next. There is also an interrupt signal with which the ACIA can interrupt the microprocessor, for instance when an external source wants to send data to the ACIA's receiving circuits to be sent on to the microprocessor.

When the microprocessor reads the Status Register of the ACIA, which it does by selecting the chip and setting some of the "register select" lines into the chip, what it is usually looking for is either a message that the Transmit Data Register is empty and needs another byte from the microprocessor, or that the Receive Register is full (has a byte that the microprocessor should read). Those messages are simply bits in the Status Register: the rightmost bit, bit 0, is called "Receive Data Register Full," and bit 1 is called "Transmit Data Register Empty."

In our case, it would be the latter that the microprocessor would use. It has to send three bytes of information to the ACIA in turn: the tens digit, the units digit, and the tenths digit. It would send the tens digit, then keep checking the Status Register until the Transmit Data Register Empty bit becomes a 1, indicating that

the tens digit has been transferred to the Transmit Shift Register, and the units digit can now
be sent to the Transmit Data
Register.

Suppose (as is actually the case)
that the printing device is rather
slow: perhaps it can print a digit
in a tenth of a second. That's
still 100,000 microseconds — and
remember that the microprocessor's basic clock can be as fast
as one microsecond. So, after the
microprocessor has sent the tens
digit, it may have to wait for
100,000 clock cycles before the
next digit can be sent. Normally,
we would program a loop that
continually reads the Status Register in the ACIA, waiting until it
comes back with the "empty"
message.

What will actually happen is that
as soon as the tens digit is sent to
the ACIA's Transmit Data Register, it will (in a matter of microseconds) be transferred to the
Transmit Shift Register, and
begin to be sent to the printer.
The Status Register will then say
that the Transmit Data Register
is empty, and the microprocessor
will send the units digit. Now
the waiting begins. The printer's

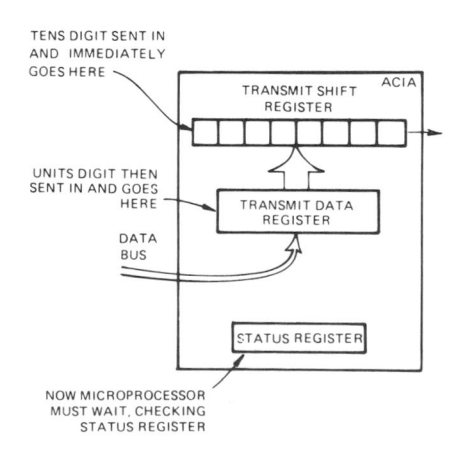

clock may be quite slow, and cause the bits to be shifted out to the printer very slowly over a tenth of a second. Or it may be fast: if there is another register in the printer, the transfer across the connecting cable can be done in a very short time. Then, however, the printer will take its tenth of a second to print the digit, and may simply hold up its clock so that no more bits are received until that has happened.

However, to make things easier for printer designers, other signals, notably two called "Request to Send" (from the ACIA to the printer) and "Clear to Send" (from the printer to the ACIA) are provided. The ACIA pays attention to these signals and doesn't attempt to send bits unless it knows that the printer (or whatever is on the other end) is ready.

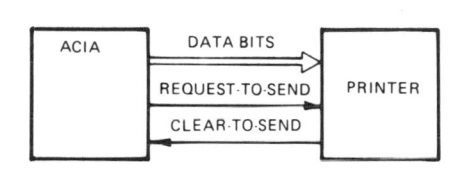

The ACIA also can check for certain error conditions. For instance, if the data being received by the ACIA has what is called a "parity bit" with each byte, the ACIA can check whether one of the bits in the byte was lost. Parity bits are commonly added to bytes so that (for instance) the sum of the 1's in a byte will be

even. Then, after the transmission across the wire, if the sum is no longer even, then a one-bit error must have occurred in transmission; the microprocessor can be told that, after which it can request the data to be retransmitted.

The ACIA can add parity bits to outgoing bytes as well as check parity bits in incoming bytes. There are several options: the sum can be requested to be even or odd (said to be "even" or "odd parity"), and the number of bits per byte can be selected to be seven or eight (both are common conventions). Through bits sent by the microprocessor into the ACIA's control register, any of these options can be selected.

Along with the byte, the ACIA either transmits, or expects to receive, either one or two "stop bits." These come at the end of the byte, and are included according to industry standards; some devices with which the ACIA would be communicating would expect one stop bit, while others would expect two. This can be selected by bits sent to the control register.

STOP BITS

1 0 111 0 11

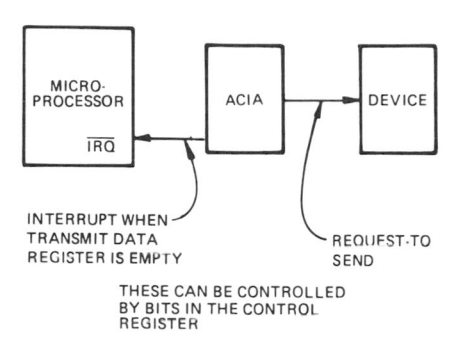

INTERRUPT WHEN
TRANSMIT DATA
REGISTER IS EMPTY

REQUEST-TO
SEND

THESE CAN BE CONTROLLED
BY BITS IN THE CONTROL
REGISTER

Among the other selections you can make with bits in the control register are the enabling or disenabling of an interrupt which can go from the ACIA to the microprocessor whenever the Transmit Data Register Empty condition arises, and the setting of the Request-To-Send line high or low (as a signal to the device on the other end).

You can also set the ACIA to expect a data bit, when receiving, either every time a clock pulse arrives from the other device, or every 16 clock pulses, or every 64. There is also a combination of control register bits that does a master reset on the ACIA, clearing all its registers.

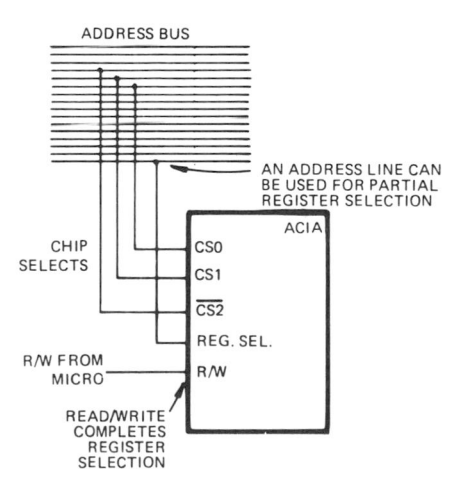

AN ADDRESS LINE CAN
BE USED FOR PARTIAL
REGISTER SELECTION

READ/WRITE
COMPLETES
REGISTER
SELECTION

To select the ACIA, there are two chip select inputs that expect high, or +5v, signals, and one that expects a low signal. To select which one of the four registers inside the ACIA the microprocessor wants to communicate with, there is first of all a "register select" line, which distinguishes the registers into two groups of two: high selects the Transmit/Receive data registers, and low selects the Control/Status registers. Singling out the

right register in each pair is done
with the Read/Write line (R/W),
which selects the read-only register or the write-only register in
each pair.

18

A DEEPER LOOK INSIDE
THE MICROPROCESSOR

In our application example, we've shown how you would program various instructions that cause the microprocessor to move data around and to modify or process it. Now let's take a little closer look at how it does so.

The data bus that goes outside the microprocessor to the rest of the world begins inside, where it connects the various registers, and the arithmetic/logic unit. As an example of how the microprocessor goes about its work, let's examine how it transfers information from one register inside itself to another.

Between each register and the data bus is a set of "transfer gates," which are really just rows of AND circuits. The AND, of course, works so that its output will go high only when both of

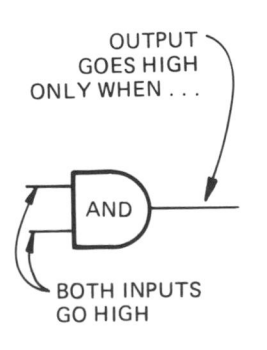

OUTPUT
GOES HIGH
ONLY WHEN . . .

AND

BOTH INPUTS
GO HIGH

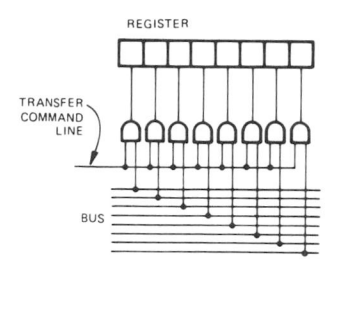

REGISTER

TRANSFER
COMMAND
LINE

BUS

its inputs are high. So to make a
set of transfer gates with ANDs,
say from the bus to a register,
one input of each AND is connected to a bus line, and the output of the AND goes to one bit
of the register. The other input
to each AND comes from the
transfer command line; when this
goes high, then the outputs of
those ANDs connected to high
lines on the bus will go high —
and set their bits in the register.

DISCONNECT
CONTROL
LINE

AND

CAUSES OUTPUT
TO BE DISCONNECTED

When going from the register to
the bus, it's a little more complicated. Here the circuits are ANDs
too, but they can also perform another trick: they can assume a
third condition, neither high nor
low, but essentially disconnected
from the bus. Then, when no
transfer is being made from this
register to the bus, the bus can
be free to carry other data.

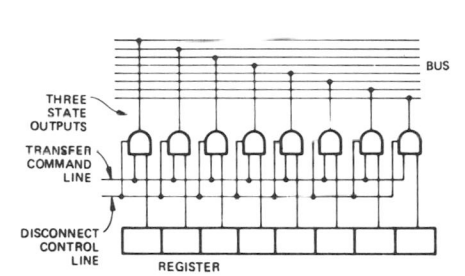

BUS

THREE
STATE
OUTPUTS

TRANSFER
COMMAND
LINE

DISCONNECT
CONTROL
LINE   REGISTER

Before a register can accept data
from another register, however,
it has to be "cleared" — have all
0's in it. A register is essentially a
row of "flip-flops" — circuits
that have a "set 1" input, a "set
0" input, and an output which
can be high or low — 1 or 0.
All that is necessary to clear a
register is to energize one line

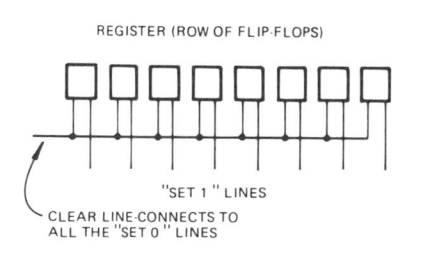

REGISTER (ROW OF FLIP-FLOPS)

"SET 1" LINES

CLEAR LINE-CONNECTS TO
ALL THE "SET 0" LINES

that goes to the "set 0" inputs of all the flip-flops in the register.

All the "set 1" inputs, of course, are connected to the outputs from the transfer gates. When some of those come on, the corresponding flip-flops in the register will be set to a "1."

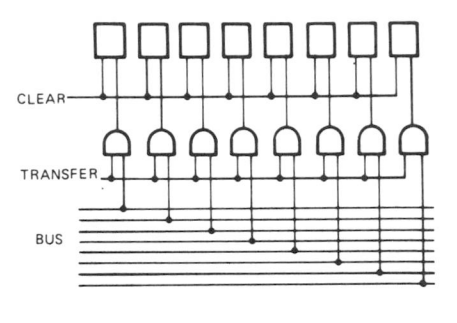

To transfer information from one register to another via the bus, then, what you have to do is first clear the receiving register; then gate the contents of the sending register onto the bus; then gate the contents of the bus into the receiving register. So long as no other signals are allowed on the bus during that time, these steps are sufficient to make the transfer.

The steps must, of course, be carried out in that definite sequence, and inside the microprocessor, that is controlled by a stored set of commands, each affecting a single control wire. For the task we have just described, a series of single control wires have to be activated (brought to +5 volts) in sequence: first, the clear line on register A has to come on momentarily; then, the

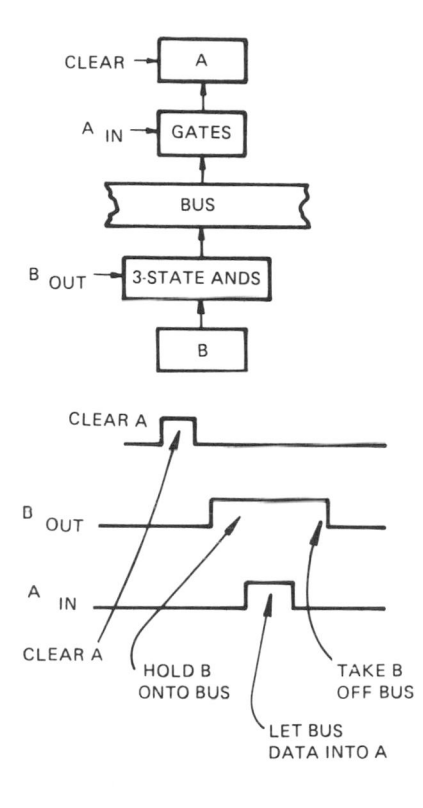

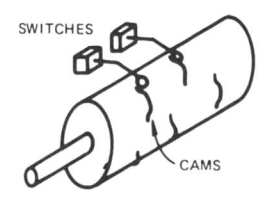

line that puts the contents of register B out on the bus has to be held on, while the line controlling the input gates to register A has to be pulsed momentarily; and finally the line that puts register B out on the bus can be "relaxed."

The way in which these operations are actually carried out, inside the microprocessor, is generally kept secret by the manufacturer; the MC6800 is no exception, as its internal design has not been published. However, we'll describe here a straightforward method they might have used, which may help clarify your thinking about the whole process that goes on in microcomputers. Remember, though, that considerably more cleverness in the design has probably been used by the manufacturer.

To make an analogy, you could think of the control lines mentioned above as being sequenced on and off by a drum controller, if the sequence did not have to be done in microseconds. The drum would turn, and bumps or cams on the drum surface would actuate switches to turn the various lines on and off. The mechan-

ism that is actually used does nothing more; it only does it faster. A small memory, of the read-only kind, is read one word at a time; in each word are bits dedicated to each control wire. As the word is read from the memory into a register, a particular bit in the register becomes a voltage on an output of the register — and is routed from there directly to the register clear line or the gating control line where it is needed. As new words are read, lines go on and off in the proper sequence.

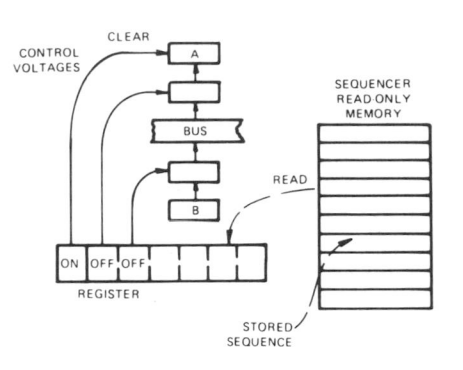

You may ask what generates the address that controls what word is read from this little read-only memory. Most likely it comes from a counter, that is, a register containing a binary number, that can be counted up by the microprocessor's clock pulses. In that way, the words in the memory can be read in a steady progression.

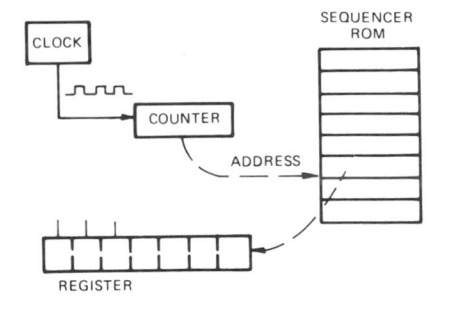

But, you may also ask, what determines whether the machine should transfer data from register B to register A, or do something else? In a real microprocessor there are many different opera-

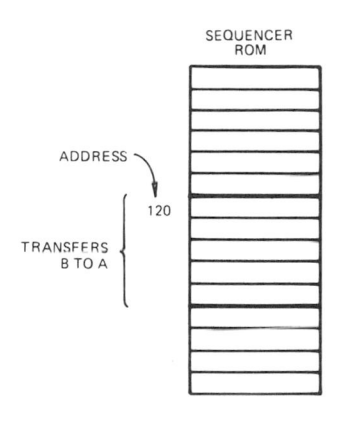

SEQUENCER ROM

ADDRESS

120

TRANSFERS
B TO A

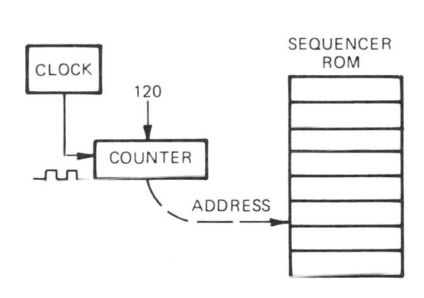

CLOCK

120

SEQUENCER
ROM

COUNTER

ADDRESS

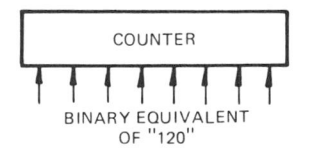

COUNTER

BINARY EQUIVALENT
OF "120"

tions like that. How does the mechanism get started on the right one?

For each such operation, there is a small dedicated part of that read-only memory. Each one starts at a particular address in that memory; for instance, the example we looked at might have started at address 120.

To get that small operation started, you have to get "120" (that is, its binary equivalent) into the counter register and then let it count; and then have some way for it to stop before it gets into the next operation.

A new number can be entered into that counter register just as a new number can be entered into an ordinary register — through input lines leading to its various bits. But where does the number come from? In most microprocessors, it is none other than the operation code of one instruction in a computer program. As instructions are read from the main memory of the microcomputer — the part of the memory where the program is kept — the first part of the instruction, the "opcode" or operation code,

is just dropped into that register. So we see that the mysterious operation codes are just addresses into the little sequencer ROM that runs all the control lines to the register clear lines and gate transfer lines, and a few other signals like those telling the arithmetic unit to shift left one bit position.

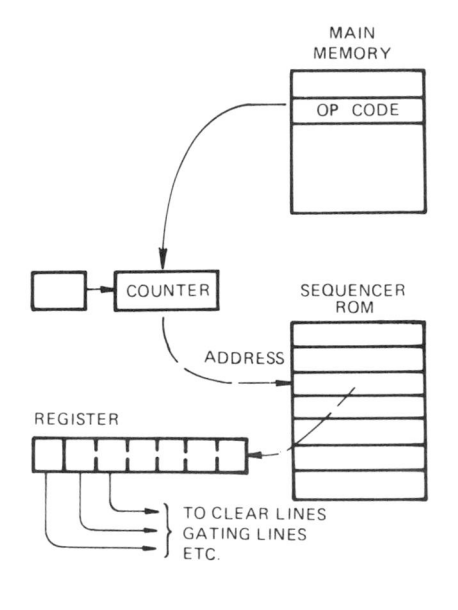

Now let's examine a more complicated operation: adding the contents of a register (let's call it "R") to the accumulator, in a typical microprocessor (not necessarily the MC6800, for which the details have not been revealed):

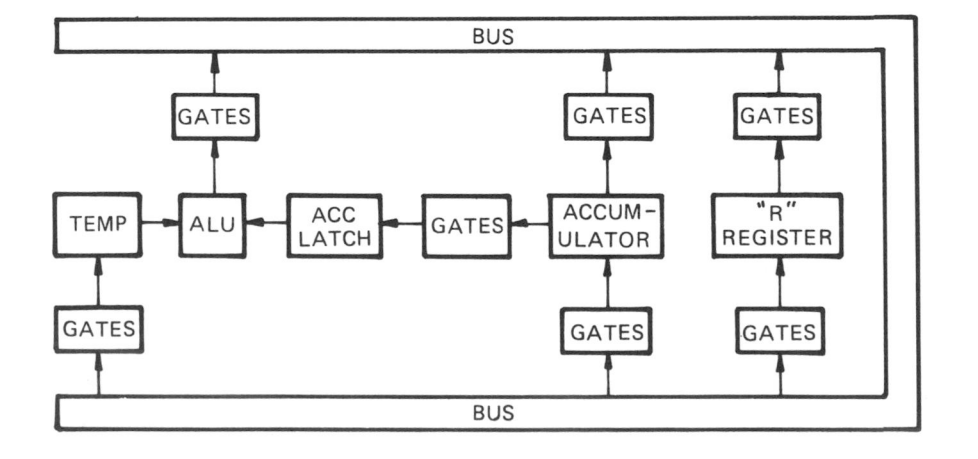

Here, the arithemetic/logic unit, or ALU, will perform the addition. Its two inputs will come from the TEMP register and the ACC LATCH register (the accumulator latch), and its outputs will go out on the bus, to be accepted back into the accumulator. The purpose of the accumulator latch register is so that the ALU will still have a good input, as its output is being sent (via the bus) back to the accumulator; it wouldn't do to have the accumulator itself be the input to the ALU while it was being changed by the ALU.

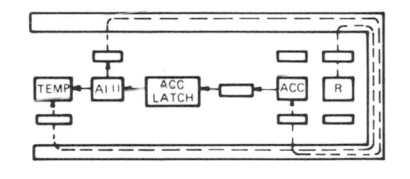

In this example, the contents of the "R" register are first sent via the bus to the TEMP register; the contents of the accumulator register are gated into the ACC LATCH register; the addition takes place in the ALU; and its outputs are gated via the bus back into the accumulator.

We can describe this more precisely with a state diagram:

When the command "Add register R to the accumulator" is re-

ceived, the operations described above are carried out, in sequence. Of course, since the gates from the accumulator to the accumulator latch are separate, and do not make use of the bus, it is possible to perform both of the first two tasks simultaneously.

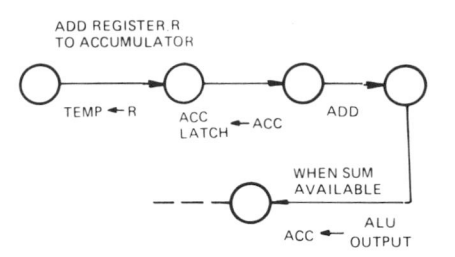

Certainly, all of these elementary operations can be commanded by single control wires, operating clear lines, transfer gate command lines, and a line to the ALU telling it to perform the addition. So the whole operation can be controlled by a sequence of control words put into the control register one by one, with the various bits turning on and off their corresponding command wires.

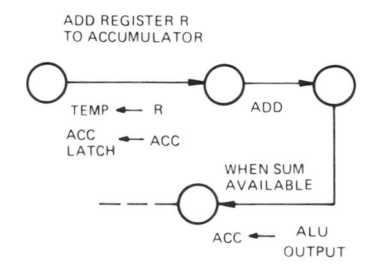

There is nothing mysterious, then, about what goes on inside the microprocessor. The only really amazing part of it is that so many different kinds of instructions are handled in a chip that has so little room for the sequencer ROM, which, of course, is right on the chip with everything else. It is not unusual for present-day microprocessors to be capable of over a hundred different kinds of instructions.

In most of this small book, we have explained the use of various features of the microprocessor through the use of our example. To make the explanation as clear as possible, we chose not to mix the complexities of computing in with the basic use of the microprocessor, its memory circuits, and its input/output circuits. In our example, we did not use, for instance, the ADD instruction; we chose a solution method which did not require computation as such. Now, however, let's look at some of these computational instructions, and see what limitations they might have in a typical microprocessor.

The MC6800 is an 8-bit microprocessor; like most others in that class, the addition and subtraction functions can be done on only 8 bits at a time. That means

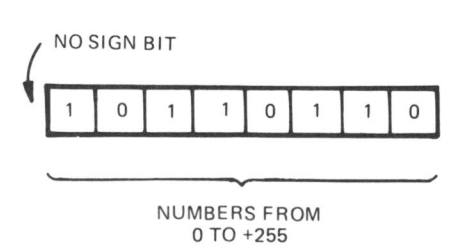

NO SIGN BIT

NUMBERS FROM
0 TO +255

that the numbers to be added or subtracted can't exceed 2^8 - 1, or 255. Not only that, but they are limited to the positive numbers from 0 to 255; a special bit for the minus sign is not included. It would seem that such a limitation would severely restrict the amount of computing that the device could do.

| 1101 0011 | 0110 0000 | 1110 0110 | 0001 1000 |

⊢ | 0000 1111 | 1001 1001 | 0000 0000 | 1111 1111 |

← | 0001 0111 |

CARRY

When the numbers to be added or subtracted exceed 255, a much more complicated set of operations has to be carried out. The numbers might, for instance, be represented as 32-bit binary numbers, each one taking up four locations in memory. That representation would allow us to cover numbers up to 4,294,967, 295. The addition, however, would have to be carried out in 8-bit pieces, starting at the right-hand end of the two numbers. Carries that would be generated at the boundaries of these 8-bit segments would have to be detected and added into the next group to the left.

An entire little program is necessary to do that. The program will become more complicated if, for example, you want to account for negative numbers. These pro-

grams are normally set up as sub-routines, which can be stored in program memory just once, then called upon to be used whenever they are needed.

You can see, then, that virtually all meaningful computation in a microprocessor has to be done with subroutines. Multiplication and division are always done with them; no instructions for those operations exist in the 6800 instruction set, nor in that of any 8-bit microprocessor.

A large and useful class of com-putations, which is also done with subroutines, is decimal arith-metic. Here, each decimal digit is represented by four bits (the "binary coded decimal," or "BCD," representation), and they are added together either one pair at a time or sometimes two pairs at a time. For applica-tions where decimal digits are coming in from the outside world, and have to be sent back (say, on displays) this might be a preferred way to do things.

Subroutines are, then, the normal mode of operation in microcom-puters; we avoided them in our example only to simplify the pre-sentation for the newcomer to

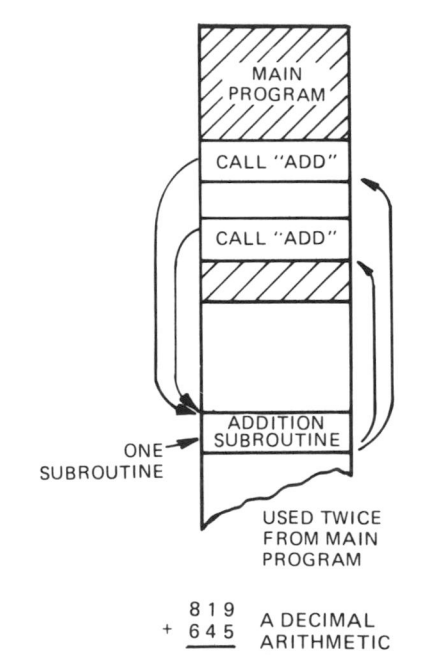

ONE SUBROUTINE

USED TWICE FROM MAIN PROGRAM

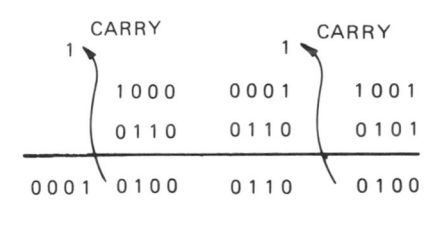

$$\begin{array}{r} 8\ 1\ 9 \\ +\ \ 6\ 4\ 5 \\ \hline 1\ 4\ 6\ 4 \end{array}$$ A DECIMAL ARITHMETIC OPERATION

SAME OPERATION IN BCD

(BINARY-CODED-DECIMAL)

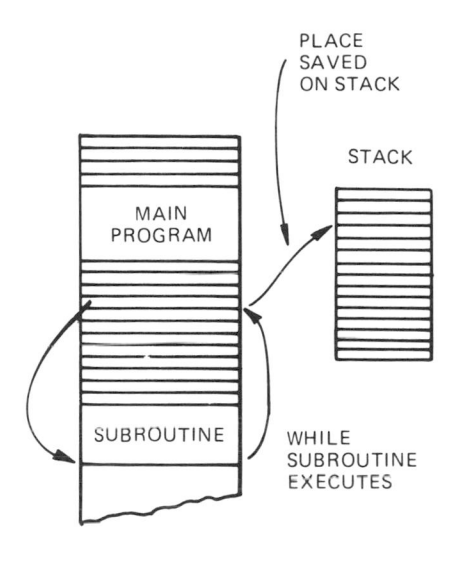

PLACE SAVED ON STACK

STACK

MAIN PROGRAM

SUBROUTINE

WHILE SUBROUTINE EXECUTES

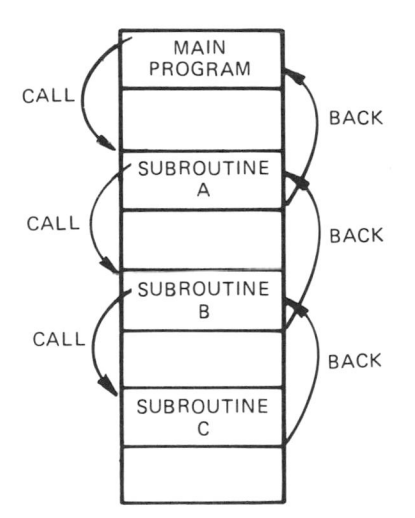

MAIN PROGRAM

CALL

BACK

SUBROUTINE A

CALL

BACK

SUBROUTINE B

CALL

BACK

SUBROUTINE C

computing. The microprocessor manufacturers, however, know full well that subroutines will be at the heart of every application, and have provided a number of special instructions to help with the process. That's the real reason for the stack that we explained in Chapter 14: since the subroutine should, for best efficiency, exist in only one place in memory yet be accessible to many different parts of the main program (or other subroutine), there had to be some way for the processor to remember where it left off, and what it was doing, before it jumped to the subroutine. That might have been done by simply reserving some extra registers inside the processor. But it's important to allow subroutines to call subroutines, and those to call still further subroutines; and then the processor will have to find its way back through several levels. The stack was the answer: each time a subroutine is called, the contents of the processor's internal registers are put on the top of the stack. If a previous set is already in the stack, the new set simply goes into the stack as well, without erasing the old ones. As the process is reversed, and the processor exits

each level of subroutine, the sets of internal register data are retrieved in reverse order — layer by layer.

Two special instructions, the BSR (branch to subroutine) and RTS (return from subroutine) instruct the microprocessor to save all these internal registers on the stack and retrieve them from the stack, all without further work on the programmer's part. The number of levels of subroutine that can be handled is limited, in the MC6800, only by the amount of memory set aside to be used as the stack.

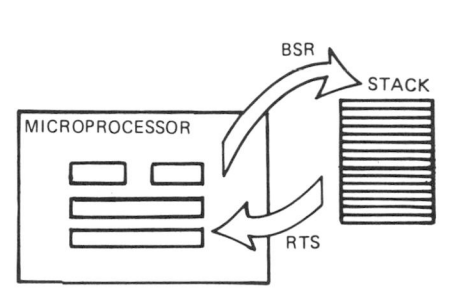

In the old mechanical calculators everyone used a number of years ago, multiplication was carried out by repeated addition. One of the two numbers to be multiplied was added to itself some number of times, depending on the rightmost digit of the other number. Then the first number, or multiplicand, was shifted left one column, and the process repeated, controlled by the next digit of the multiplier — and so on. Multiplication in microprocessors is done in basically the same way: by adding and shifting. That requires an instruction that will shift the bits in the ac-

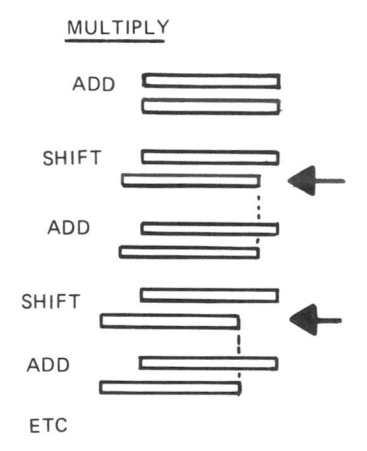

cumulator left, for multiplying, or right, for dividing.

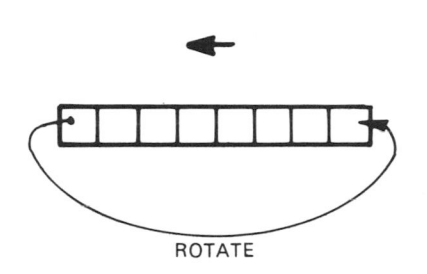

ROTATE

The manipulation of special codes, used to communicate with certain external equipment, sometimes requires a slight variation of this shifting, in which the bits that are "shifted off the end" are captured and put back on the other end; hence the name "rotate." Rotating can also be done to the left or to the right.

Often data will arrive at a microcomputer in the form of short strings of bits that are "packed" together into a standard 8-bit byte. For instance, a byte might contain two four-bit BCD characters. Often it is convenient to process these characters independently, so it is necessary to strip one of them out of the incoming byte. The logical AND instruction provides a way to do this: by ANDing a second word, called a "mask," with the accumulator, those bits in the accumulator that correspond in position to any 0's in the mask are made 0's themselves. If we had the byte "0111 0011" in the accumulator, and ANDed it with the mask "0000 1111," then

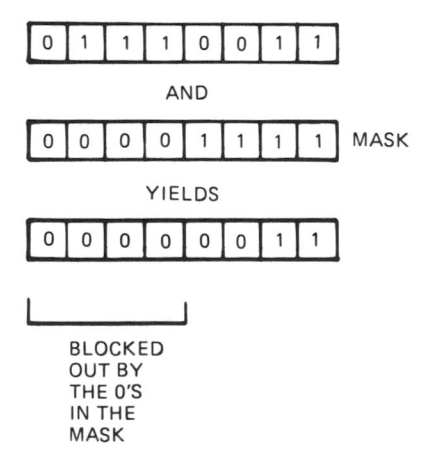

AND

MASK

YIELDS

BLOCKED
OUT BY
THE 0'S
IN THE
MASK

what would be left in the accumulator would be "0000 0011" — that is, the first BCD character has been stripped out.

The opposite operation of packing two BCD characters into one 8-bit byte can be done using the logical OR instruction, which puts 1's into the final target byte in certain positions, if either input byte had a 1 in that position. If, for instance, we had "0000 0011" in the accumulator, and "0111 0000" in a memory location, we could OR them together to obtain "0111 0011."

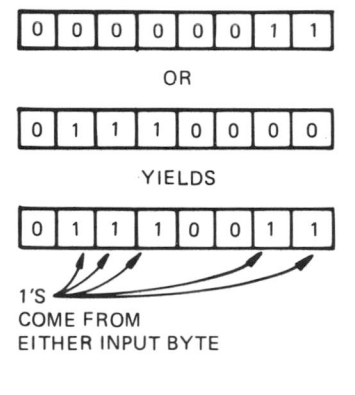

Other logical instructions are used to exclusive-OR two bytes together (the target byte gets a 1 in a certain position if it came from one input byte or the other, but not both), or to complement a byte (put a 0 for every 1, and a 1 for every 0).

In our example we used a branch instruction, BEQ, or branch if the accumulator is equal to zero. Other branch instructions work if the accumulator is greater than zero, greater than or equal to zero, not equal to zero, equal to or less than zero, or simply less than zero. There is also a COMPARE instruction that allows the

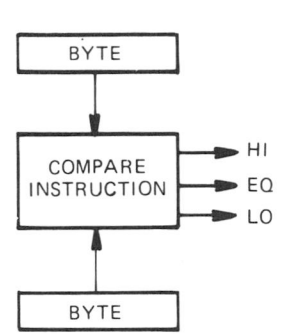

programmer to compare two bytes, one in an accumulator and one in a memory location. Then there are branch instructions that work if the one in memory is higher, lower, or equal to the one in the accumulator. Other branch instructions work on the other bits in the condition code, or flag, register: the bit that gets set during an overflow; the bit that gets set when a carry occurs; the bit that gets set when a number tries to go negative; and so on.

Clearly, the subroutines which carry out the mathematical work must use these branch instructions, to direct the computational work depending on whether such things as carries and overflows occur. The manner in which those features are used has been the subject of a great deal of work over the years: work which has resulted in algorithms (methods) for addition, subtraction, multiplication, and division of both full binary numbers (of various lengths — 8, 16, 32 bits and so on) and binary-coded-decimal numbers. These are, in turn, used in still more complex algorithms for the calculation of trigonometric functions, logarithmic functions, and so on. We will not cover these further here except

to say that the programmer should make extensive inquiries into the previous existence of subroutines like these before spending the work to re-invent them.

The instructions provided with the MC6800 microprocessor work in several "addressing modes," as they are called: there is the immediate mode, in which the data to be used by the instruction is found right in the program memory, directly after the instruction itself; then there is the "direct" mode, in which the byte right after the instruction is the address of the data, except the data has to be in the first 256 locations of memory; then there is the "extended" mode, in which two bytes (a full 16 bits) of address are carried after the instruction, and this address points to data that can be anywhere in 65,536 memory locations; and finally there is the "indexed" mode, in which the byte following the instruction in program memory will be added to whatever 16-bit number is in the index register, and the result used as the address of the data. Not all the instructions work with all those modes, but

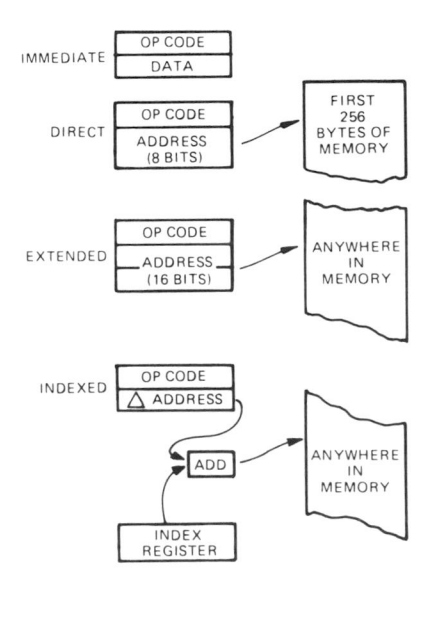

many of them work with several
of them.

Counting the combinations of
types of instructions with the
different addressing modes in
which they will work, there are
197 different combinations in
the MC6800. The operation code
(the first byte of the instruction)
is different for each of these.
That is all the more remarkable
when we remember that all of
these instructions are handled —
recognized, decoded, and acted
upon — within a single microelectronic chip.

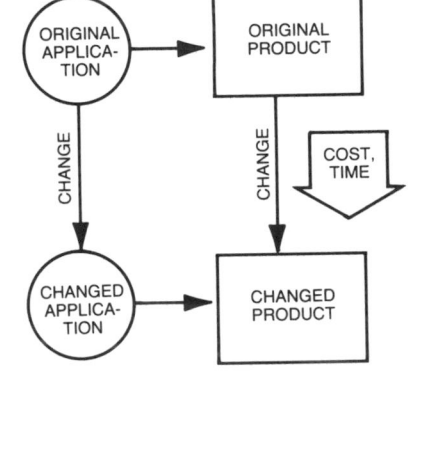

So far, we have shown software, written in assembler language, that tries to be fairly efficient in its use of memory in order to keep the hardware cost of the application as low as possible. In many applications, however, that concern may not be the most important one. Memory is much less expensive now than it was a few years ago. What if you are concerned that some aspect of your application will change, and you want to be sure that you can adapt your product or system to the change as rapidly as possible? That might mean more to the economic success of your business that would a small price reduction resulting from a slightly smaller memory.

This situation is compounded when the application is very complex. The example we have used in this book is tiny compared to

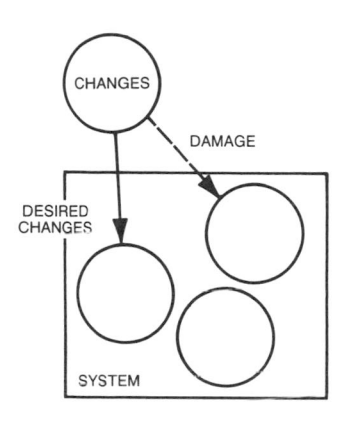

some applications that exist in industry, especially in the defense industry. When changes have to be made in a really complex software system, it is very difficult to change one part of the software without damaging some other part. Then the cost of the change must include the cost to fix that damage, as well as to fix any havoc that the damaged software might cause when the application is run. In the case of aircraft software, for instance, the damaged part might result in lost lives. Controlling software changes, in the face of great complexity, is one of the most difficult modern challenges in software.

Three major thrusts are taking place in the software world today that are intended to deal with this problem. The first is the development of new languages designed to make software more understandable (to someone who has to change a program) and more resistant to damage from wrong or incomplete changes. The U.S. Department of Defense had those goals when it commissioned the "Ada" software language, intended for very complex military applications.

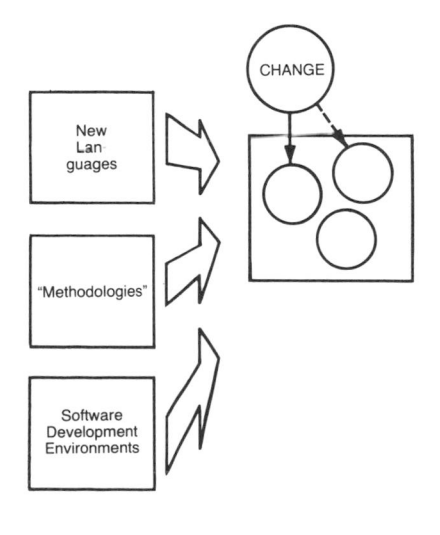

The second major thrust is the development of new software development methods (or "methodologies," as they are sometimes called) that help to partition the software and package it so that when a change comes in, it is likely to affect only one partition — and not make its presence felt in dozens of different places.

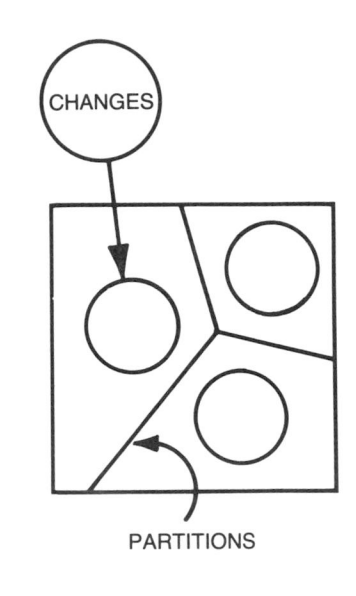

PARTITIONS

The third major thrust is the development of software development "environments," or collections of hardware and software tools, that help in all stages of software development, including keeping track of the changes and all the various versions and revisions that result For lack of such an "environment," some large software systems have been known to collapse completely—to have had so many changes added that nothing worked anymore, and to have had no way to back the changes out, one at a time, until the software worked again. Those parts of the "environment" that handle changes and versions are sometimes called "configuration of management tools."

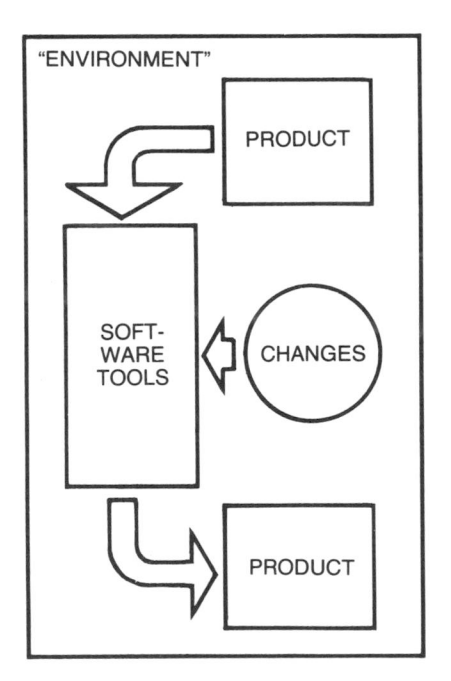

21 THE ADA LANGUAGE

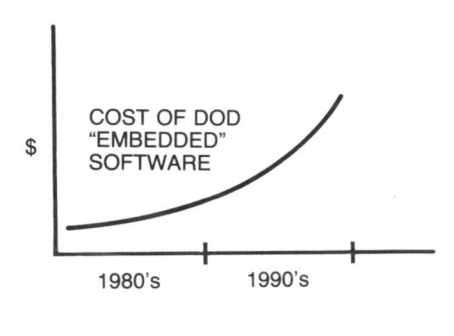

COST OF DOD
"EMBEDDED"
SOFTWARE

$

1980's   1990's

Ada, an important new language, was commissioned to be developed for the U.S. Department of Defense. DoD has been projecting for a number of years that its total cost for "embedded" software (software that runs in computers embedded in aircraft, ships, tanks, and so on) would be more than 30 billion dollars a year in the early 1990s and would continuously increase. In addition, some 70% of that cost would be for "software maintenance" — that is, changes. Most of those changes would not be to fix errors — they would instead be due to changing application requirements.

To help solve this problem, DoD had the Ada language developed in the late 1970s and early 1980s. Ada tries, first to make computer programs more understandable and readable so that when a change is necessary, someone

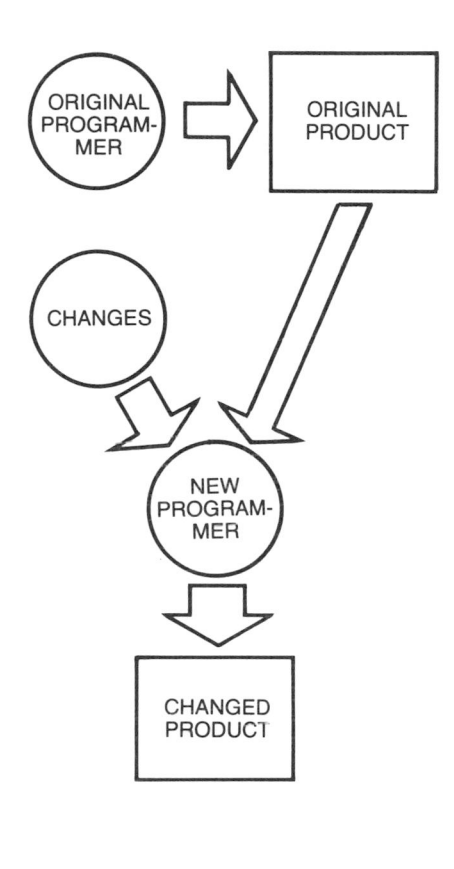

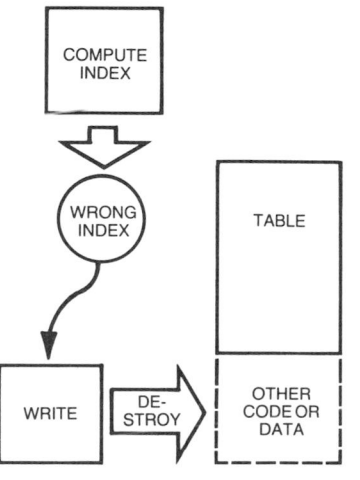

other than the original programmer can understand what it does. For instance, long names are allowed for data elements — names that can be whole phrases, such as "INTEGRATOR_RESET_ AND_START_SIGNAL." If you are working in a commercial software development business, you might elect not to use the Ada language — it is a very complex language to learn — but in choosing a language, you should select one that allows long, understandable names for data elements.

Then, Ada has facilities with which a programmer can protect his code from accidental damage or erasure during computer execution — something that can very easily happen with the assembly language we used in the first part of this book, or even with higher-level languages like Fortran. For example, in those languages, if you compute an index to be used to access a table and then write something into that table, there is nothing to prevent you from computing too large an index value and thereby writing over something that is outside your table — and perhaps destroying some other programmer's data or executable code. Ada has a way to

erect "fences" that prevent this from occurring and so limit the damage a bad change can do. Ada checks for problems of this kind during the compilation phase and then builds in extra software that makes similar checks during actual execution. Some other general-purpose languages have those features, but they are not suitable for embedded systems.

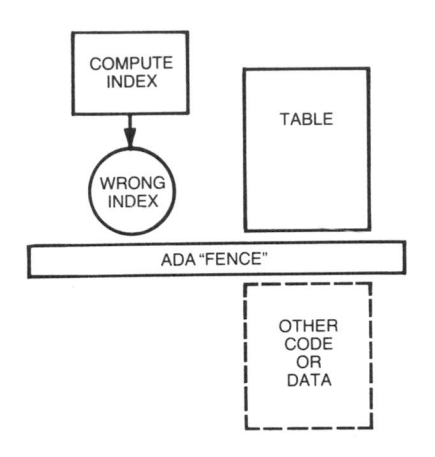

Ada also has features that allow the designer of the software to check out the "interfaces" among the different pieces of his design — that is, data elements that flow from one piece to another. Special characteristics can be described for each data element, such as whether it is an integer or has a decimal part, or perhaps is a collection of simpler data elements. Then — before the programmer writes any executable code — Ada will check those descriptions against the data element names in all the different pieces of software and flag all the cases where the descriptions for the same name don't match. In a very large programming project, where each piece of software is being developed by a different person, this can prevent mismatches and misunderstandings. Later, when the

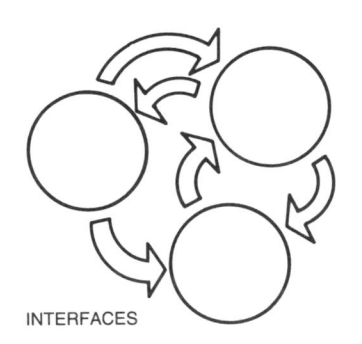

INTERFACES

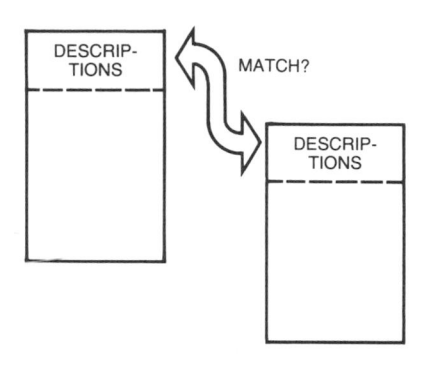

changes come in, Ada will do this check again, to make sure that a changed piece of software still matches its neighbors in its data element descriptions. In assembly language or Fortran, if you had these kinds of errors, nothing would catch them — the program would simply give bad results, and then you would have to search for the reason. There are simpler languages than Ada that have some of these features, but Ada is the only language so far that has been designed specifically to solve as many of these problems of changing software as possible. Although you may not, for good reasons, want to program your applications in Ada, you probably will want to pay attention to these problems and make sure you have considered them fully. So we have chosen Ada to use as the example for this section of the book, since we believe Ada demonstrates these principles better than other languages.

22

SOFTWARE DEVELOPMENT METHODOLOGIES—AND OBJECT-ORIENTED DESIGN

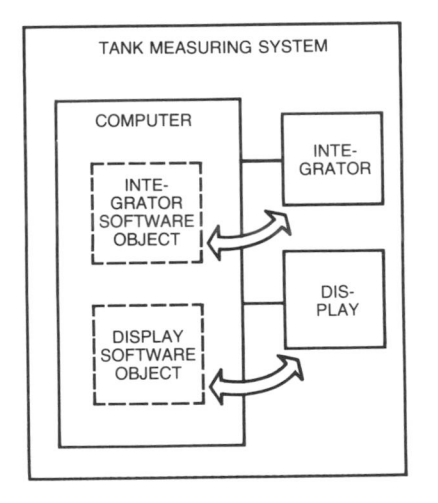

Of the new software development methods or "methodologies," perhaps the one that does the most to help "bulletproof" software against bad changes is one called "object-oriented design." (Software methods tend to come in many different varieties, and that is true of object-oriented design too. We'll describe what we believe to be a form that is usable with most general-purpose languages.) In this method, the software is divided into "objects," or blocks, which relate, if possible, to a single piece of hardware (or other entity) in your application. For instance, in the example in this book, one software object might be created to deal with the integrator and another to deal with the display. Then if the integrator or the display had to change (let's say you wanted to switch to a much cheaper display for future sales),

the changes to the software could
be limited to that one object.

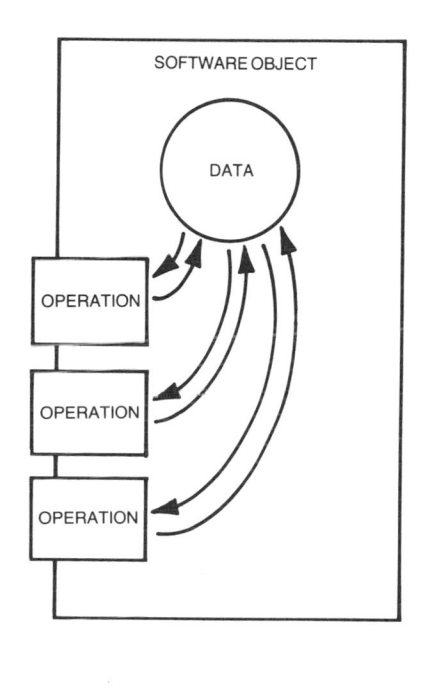

Software objects usually consist
of data (such as our table of digit
codes versus the time count) plus
some executable code, and typically the code will be divided into
"operations" that each operate on
that data in some well-defined,
easily understood way. That further divides things up so that a
new programmer, coming in to
make a change, can easily find
just the places that need to be
changed and won't go changing
(and damaging) lots of unrelated
code.

23

OBJECT-ORIENTED PARTITIONING, WITH ENTITY-RELATIONSHIP DIAGRAMS

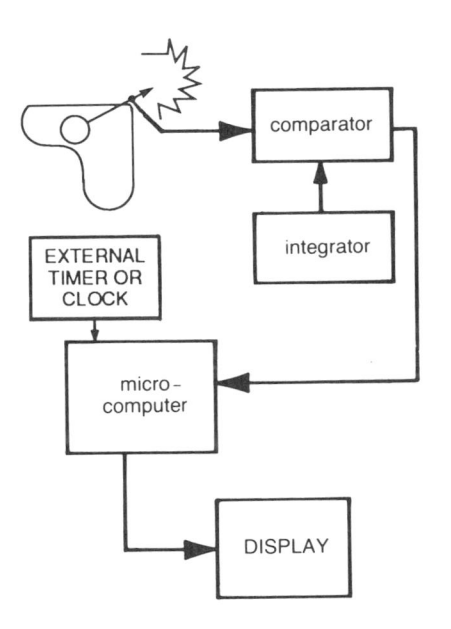

There are any number of ways you can partition your software into "objects" that satisfy these objectives. The method we will describe begins with a diagram of the computer and all of the external hardware involved in the application — essentially the same diagram as the one on page 39. However, there is one big difference: instead of using the computer itself as the timing device, we will use an external electronic timer, which produces periodic signals — "clock ticks" — that are a definite number of microseconds apart. By counting these with the computer, we can determine how long the integrator has been operating before the comparator fires. We'll need to feed the output of the external interval timer (which we'll call the "clock") into the computer on an interrupt line — and we'll discuss that at length a little later.

Next, using this diagram as a guide, we draw what is called an "entity-relationship" diagram. Again, these come in different flavors, but here is our style. To create one, you first make a drawing with a block for each of the hardware elements in your application. In our example, you would have included blocks for the tank, the float, the potentiometer, the integrator, the comparator, the clock, the displays, and the computer. We say "hardware elements," but you can include the user — he or she is an entity.

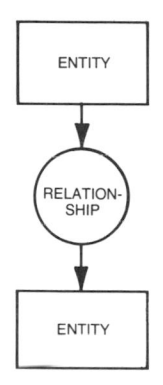

Then you would draw, between them, some round shapes representing "relationships" — any possible way the entities might be related, expressed usually as the predicate of a sentence.

For instance, the first entity might be "a tank of a certain size and shape." The second entity might be the float. Because the size and shape of the tank "affects the motion of" the float, that becomes the relationship. Another entity, "liquid in the tank of a certain amount," actually "moves" the float, which in turn "moves" the potentiometer. In this way, all the parts of your application

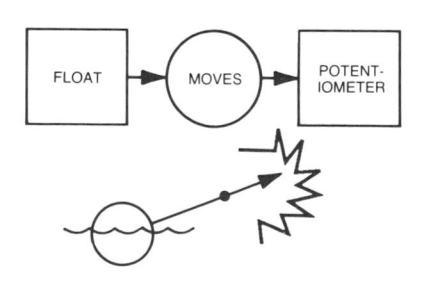

are named and related to all the
other parts.

There is one unusual entity we can
show in our diagram: "time." We
can show this to emphasize the
connection between the integrator and the clock. Time drives
them both. So an entity can in-
deed be anything that has to do
with the application and, when
mentioned, adds to our understanding of the problem.

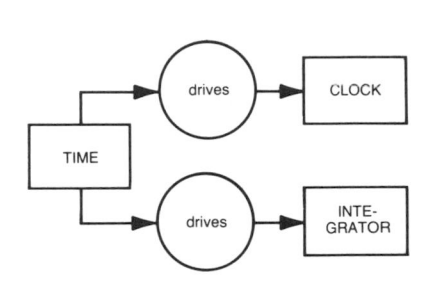

When you come to the computer,
typically you will find that it is
related to a number of different
entities in the system, with some
complicated relationships.

The computer, however, performs
a number of different jobs, and so
can be considered (for the purposes of this diagram) as a number
of different entities. A convenient
way to think of this is to consider
the computer plus the first piece
of software to be one entity, the
computer plus the second piece of
software as a second entity, and
so on. Now you are beginning to
partition the software, and you
have to consider how to do that.

To make software objects that
will really help when it is time to
make changes, you should first

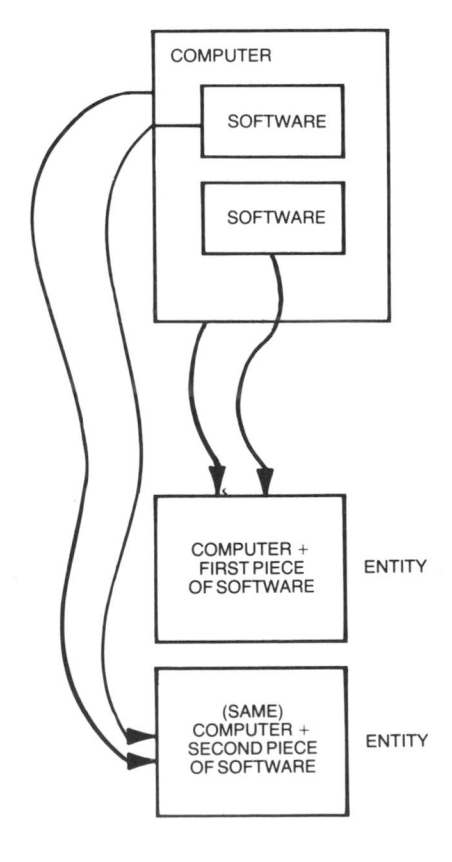

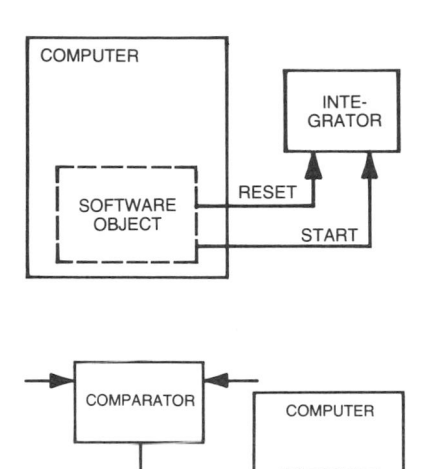

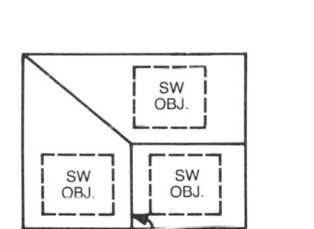

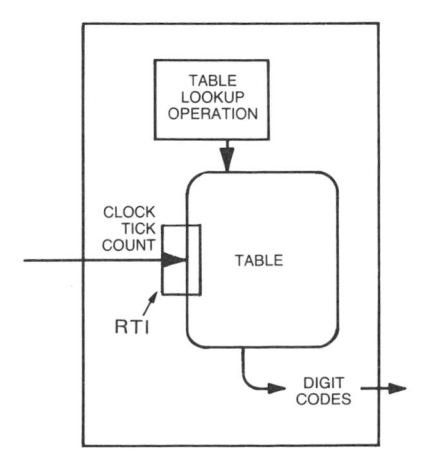

make one software entity that takes care of, or services, each of the noncomputer entities in your application. For example, the integrator needs to be reset to zero volts and started up. You can set up one software entity, or object, that does that. The comparator emits a signal — which says the integrator voltage now equals that of the potentiometer — and the computer needs a software object that listens for, and responds to, that event. The "clock" or interval timer emits periodic clock ticks, and a software object can listen for them via the *run time interface* (RTI) described on page 195. The displays need to be fed with signals representing the digit codes, so you need a software object that will do that. In setting things up this way, you will be partitioning the software so that if a change to any of those external hardware elements comes in, only one piece of software will be affected.

Of course, there will be other software besides these pieces. In our case, we might include the process of counting the clock ticks in the program that deals with the clock. However, the program that will convert the clock tick count to the digit codes for the display should really not go into either

the program that deals with the clock or the program that deals with the displays. Here there is a large data structure required to represent the table of digit codes versus clock tick count — and there should be a separate part of the software that deals with this data structure and protects it.

Finally, there should be a control program that sequences all the other modules, determines when they have completed their individual work, and transfers data from one to another. In that way, none of the other five programs will need to "know" any details about any of the others. All that kind of knowledge, and the knowledge of how to sequence the others to produce the overall solution, is distilled into one place, where it can be found easily and where it will not be mixed up with issues of dealing with data structures or the outside world.

Now the relationships among these six programs can be filled in, and they will include the actions of transferring the data around, starting and stopping the other programs, and asking the other programs for notice of critical events, like the "equals" signal from the comparator.

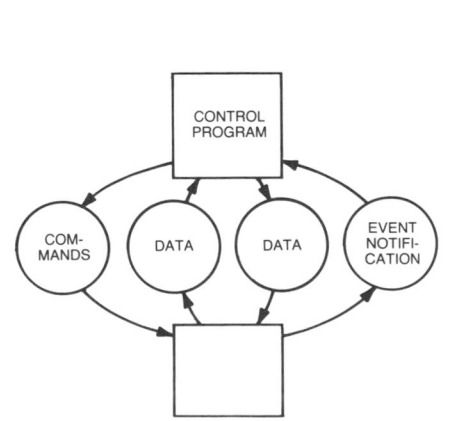

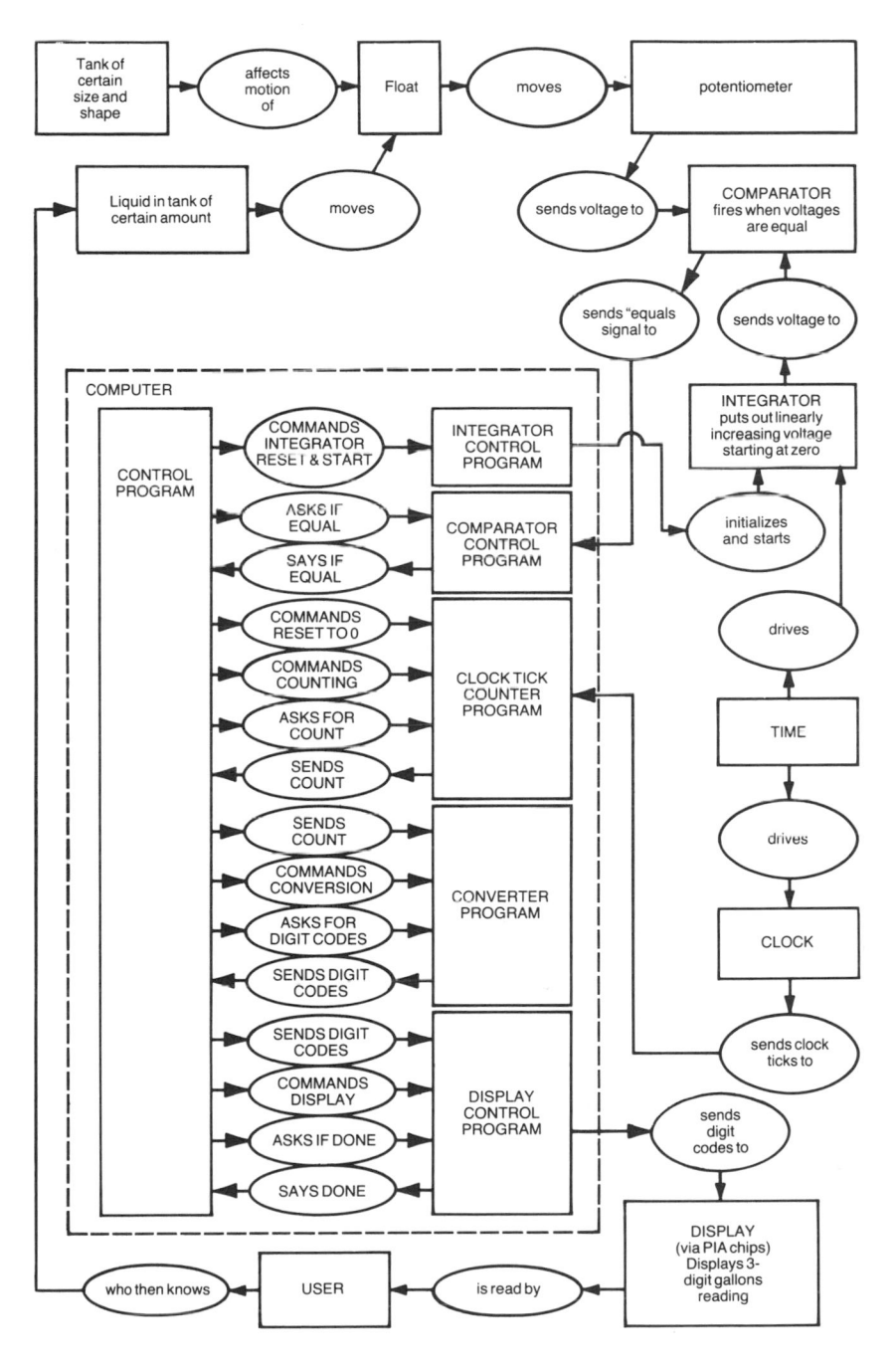

ENTITY-RELATIONSHIP DIAGRAM EXPANDED TO SHOW SOFTWARE OBJECTS

24   THE STRUCTURE OF ADA

We said earlier that Ada has been designed to keep various aspects of the program separated, so that if a requirement for a change comes in, it will usually strike only one place in the code. But exactly what structure does Ada use to do this?

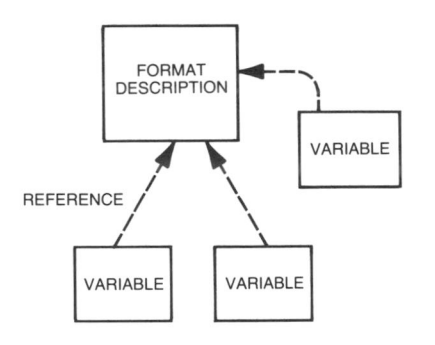

First, there is the description of data items. Ada recognizes that you may have a number of data items that are identical in format. You would like to describe the format in one place, and then, for each instance of its use, just reference that format description. Then, if the modification involved a change of format that affected all the instances equally, you would only have to change the format in that one place.

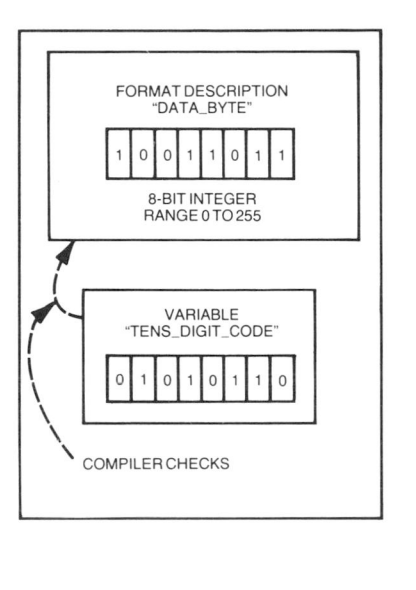

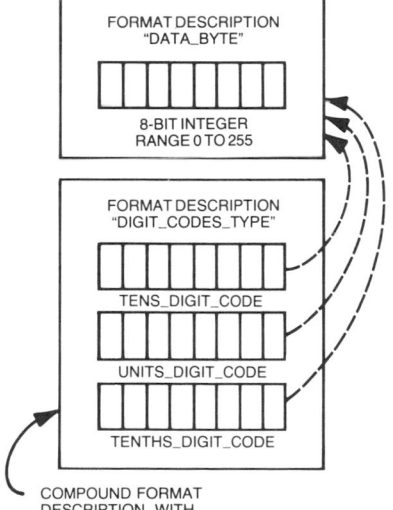

COMPOUND FORMAT
DESCRIPTION, WITH
PARTS REFERENCING
ANOTHER FORMAT
DESCRIPTION

To accomplish that, Ada provides for "type" definitions, which are the format descriptions. For instance, if we wanted to define a format for the digit codes we need to send to the display, we might say:

type DATA_BYTE is integer range 0..255;

for DATA_BYTE size use 8;

This defines the format as one eight-bit byte. Then we could define one of the three digit codes we need to use as:

TENS_DIGIT_CODE: DATA_BYTE;

— and Ada would then check this variable, each time it was updated to a new value, to see that it remained within the format definition of a single eight-bit byte.

However, the format can be even more complicated. You could define a type called "DIGIT_ CODES_TYPE" that defines a group of three-digit codes all together:

type DIGIT_CODES_TYPE is record
   TENS_DIGIT_CODE: DATA_BYTE:
   UNITS_DIGIT_CODE: DATA_BYTE;
   TENTHS_DIGIT_CODE: DATA_BYTE;
end record;

This "record," then, is a group of three eight-bit bytes.

For instance, now you can define two different variables as being of this type:

    DIGIT_CODES: DIGIT_CODE_TYPE;
    NUMERIC_READOUT: DIGIT_CODES_TYPE;

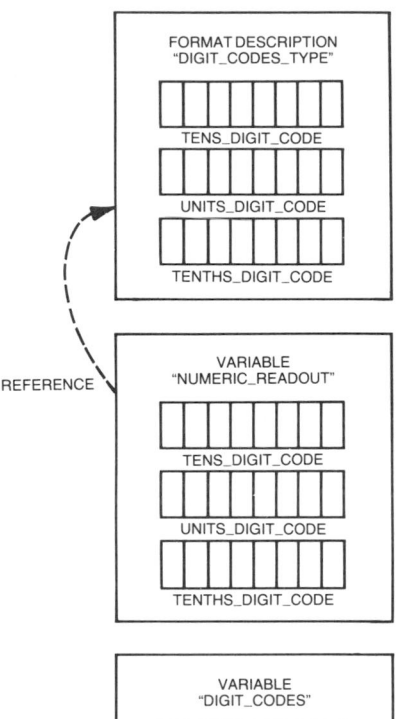

And you could transfer the contents of one to the other, if you needed to:

    DIGIT_CODES:=NUMERIC_READOUT;

— and all three subparts of the format — all three eight-bit bytes — would be automatically transferred without specifying them individually (if that's what you needed to do).

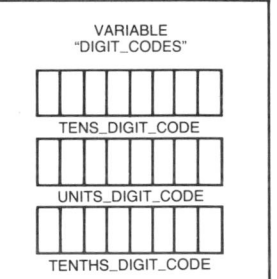

If the format changes, due say to a change in system requirements, you would just change the definition of "DIGIT_CODES_TYPE"; and the definitions of "DIGIT_CODES" and "NUMERIC_READOUT" would change automatically. There is only one place in the program where you have to do that, and inconsistencies are prevented.

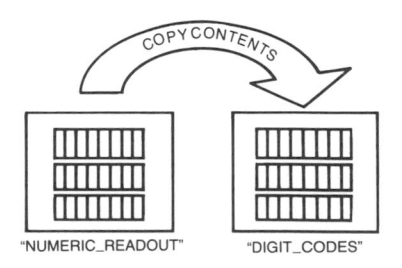

The use of these "types" as generic definitions is part of a property of Ada called "strong typing." It does exist in some other modern, general-purpose programming languages, like PASCAL, though in some respects the "typing" is not as strong. Strong typing is one of the design features that makes Ada what it is, but there are others.

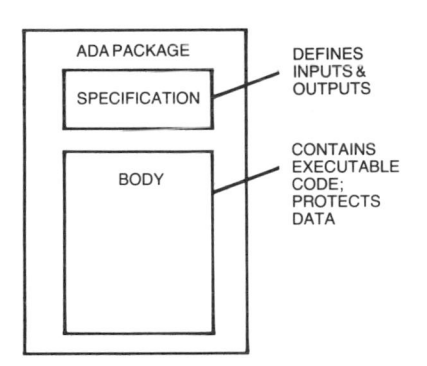

Probably the most important is the idea of the "package": that is, Ada's way of dividing up programs so that each is protected from changes made to the others. The package concept is based on the idea that it should be possible to isolate the definition of the inputs and outputs of a section of the program from that section's executable code, so that the programmers of the other sections don't need to concern themselves with the code (and are thereby prevented from causing harm to the code, albeit unintentionally). The definition of the inputs and outputs is done in the package "specification," and the executable code is put into what is called the "package body." We will show how our tank readout example makes use of this fea-

ture. No other language has quite this mechanism for separation and mutual protection — but you can see how important it would be in a very large software job where several hundred people are programming simultaneously.

Ada has several other features that promote this separation and resistance to inconsistency in the face of change. Those features require Ada to have a complex grammar or syntax, which in itself causes Ada to be harder to read and figure out — which isn't what you would want. That effect is balanced by the longer, more meaningful names for variables (data items) and procedures (pieces of executable code), and the result can be quite readable if the programmer takes the trouble to use the longer names and tries to make them meaningful. Still, programmers must get used to the more complicated grammar and syntax, and on a large project with many people new to Ada, that's an extra, up-front cost. But after they do learn, the benefits of consistency make themselves apparent.

25

OUR EXAMPLE IN ADA—
A STEP-BY-STEP
EXPLANATION

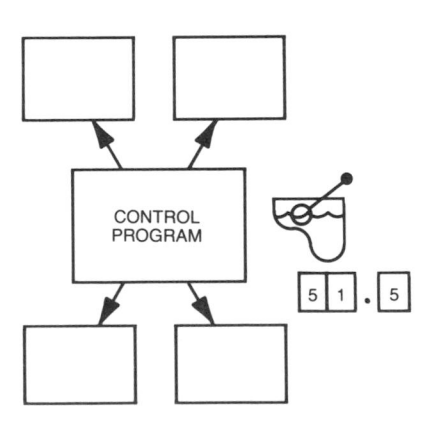

We have seen that Ada allows us to use long names for variables, constants, and segments of executable code, which helps to make the Ada code readable and understandable. However, as we said, Ada also tries to break the various facets of the code down into pieces so that a change to the problem will usually strike only one part of the code, and the result is that Ada syntax is not especially easy to read. Because of that, we won't explain the Ada program for the tank measurement system just by starting at the beginning of the program. Instead, we'll begin with the middle of the control program, because that's the part that corresponds most directly to the tank measurement problem; then, we'll go on to some of the other parts shown on the entity-relationship diagram. We'll show examples of the parts

189

of the code that have been broken
out so that changes can be made
more easily, and why each of
those parts is there. This won't be
a complete guide to Ada programming — that would take a whole
book by itself — but we'll explain
all the important features we use
in the tank measurement problem. As you read this explanation, you may want to refer to the
complete listing of the Ada program. It's given in Chapter 34.

26

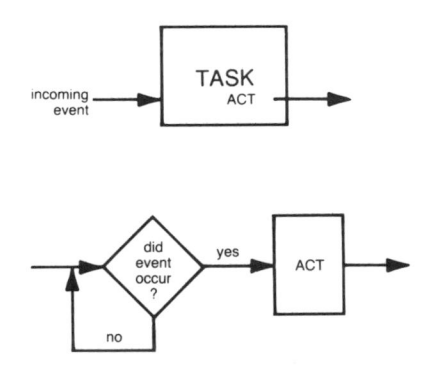

In the middle of the control program is something that in Ada is called a "task." A task is a piece of code that is able to wait to accept messages from some other piece of code or an outside event, and then do something when that happens. The central part of this task starts this way:

```
START_OR_RESET: loop
NORMAL_OPERATION: loop
```

This sets up two loops, one inside the other; while things are operating normally, we'll be inside the inner one, but when we need to start or restart the operation, we'll be in the outer loop. The program continues:

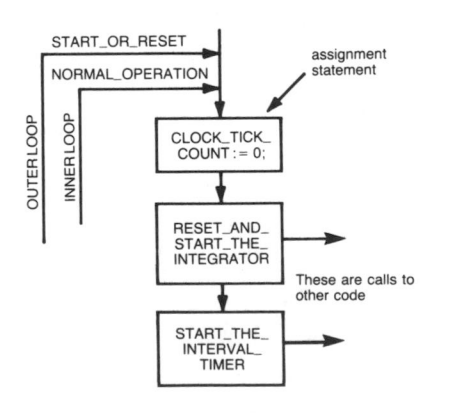

```
CLOCK_TICK_COUNT:=0;
```

which sets the clock tick count to zero. The ":=" is the assignment symbol; it places zero in the memory location named CLOCK

_TICK_COUNT. Ada statements end with a semicolon. Then:

RESET_AND_START_THE_INTEGRATOR;

In Ada, calling or activating another piece of code — a subprogram or procedure, or what we have been calling a subroutine in assembly language — is done just by mentioning the name of that routine, which here is "RESET_AND_START_THE_INTEGRATOR." Because the name can be so long, it can be self-explanatory. The program continues with another call:

START_THE_INTERVAL_TIMER:

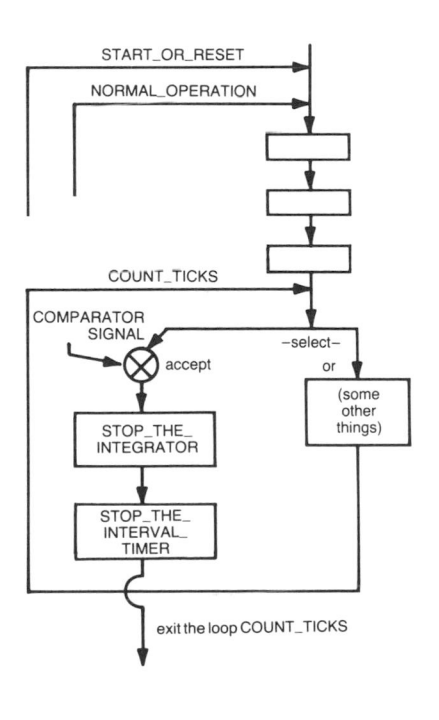

Next, we have a loop called "COUNT_TICKS," which is inside the loop "NORMAL_OPERATION." The following is only the first part of the "COUNT_TICKS" loop:

```
COUNT_TICKS: loop
select
  accept COMPARATOR_SIGNAL;
  STOP_THE_INTEGRATOR;
  STOP_THE_INTERVAL_TIMER;
  exit COUNT_TICKS;
```

Despite the unfamiliar syntax, you've probably figured out just what this does: the system is going to "accept" a signal from outside (in this case, the comparator signal). "Accepting" the signal just

means that at this point in the program the task is going to look to see if the signal is there. If it is, then we are going to stop the integrator and the interval timer and exit out of the loop COUNT _TICKS, to a place where we will do the conversion of the count to the digit codes. But what if the signal is not there? We have to provide code to take care of that case, and that's what the "select" in the above Ada code means: it says that we either select to accept the comparator signal, or, if it's not there, then do something else.

Now the comparator signal is, of course, an electronic signal, and in Chapter 13 we described how it is connected to the interrupt input on one of the peripheral interface adapters and from there to the interrupt input on the microprocessor. Then, as we described in Chapter 14, we wrote some code that "handled" the interrupt — code that will be executed when the interrupt arrives. Before we go on with our present discussion, we need to stop and consider how interrupts are handled in Ada, because we'll need to know that — not only for the comparator but for the external timer or "clock."

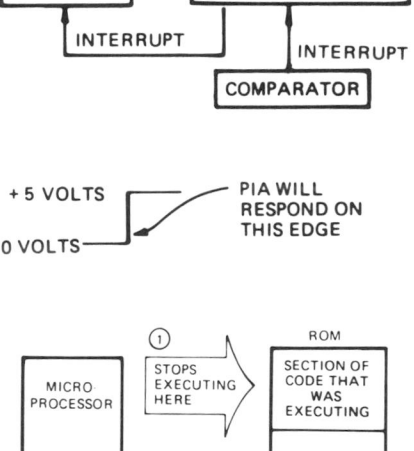

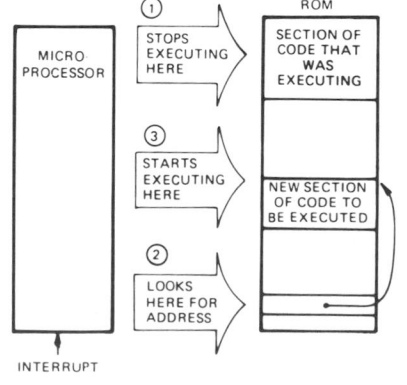

27

INTERRUPTS IN ADA

In talking about how interrupts are done in Ada, we have to remember first that a large part of the interrupt handling mechanism is not really part of the Ada language as defined by the U.S. Department of Defense. Instead, it's buried in code that is either built by the compiler or supplied in a "run-time support package" designed for the microcomputer you are using. Either way, this code is what's called "implementation dependent," meaning that each supplier gets to do it his own way, with no standardization. Here, we'll describe one of the ways, but you must remember that other ways are not only possible, but common.

First, although we will be using a modern microprocessor and perhaps a complete, ready-to-go computer-on-a-board, let's assume

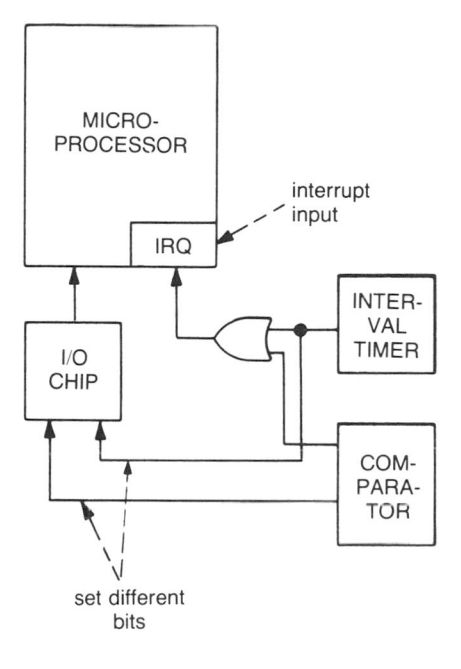

set different
bits

memory location
dedicated to
interrupt

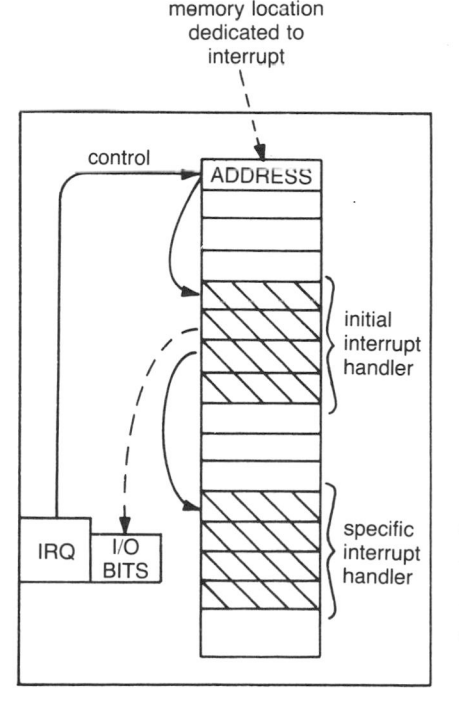

that we have the same situation we had with the MC 6800: that there is only one interrupt line into the microprocessor. If we need two separate interrupts — and here we do, one for the comparator signal, and one for the clock ticks coming from the interval timer — then each must work by setting a bit in an I/O chip to say which interrupt it is and then "ringing the doorbell" by pulsing the interrupt line.

When that interrupt comes in, the place where the microprocessor is currently executing will be saved, and control will be transferred to whatever address is stored in the special memory location dedicated to the interrupt line. The code that is located there will first have to interrogate the I/O chip bits, in order to tell which interrupt it was. Then it can (somehow) transfer control to a handler routine for that specific interrupt.

Let's call the piece of code that does those initial steps the "initial interrupt handler" — we'll come back to it presently. Now we can look at what happens in Ada.

To explain how interrupts are
handled in Ada, we first need
to say some more about Ada
"tasks," which are used to respond to interrupts as well as to
many other things.

To expand the brief definition we
gave above, "tasks" are pieces of
code that you can think of as being executed simultaneously and
in parallel, and which can wait for
one another in order to exchange
information. Of course, if the various Ada tasks are all operating
in the same computer, then they
can't really work simultaneously.
But the Ada code that is generated
by the compiler will switch the
execution back and forth among
the tasks in order to simulate simultaneous action. The way it
does this varies with the implementation, but if it is a "time-
slicing" method, then every so
many microseconds this switch
will take place.

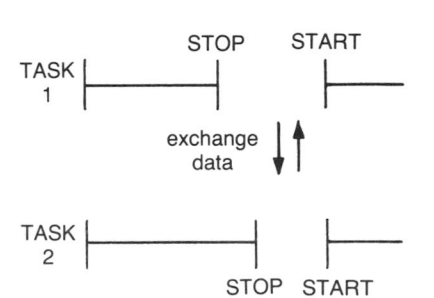

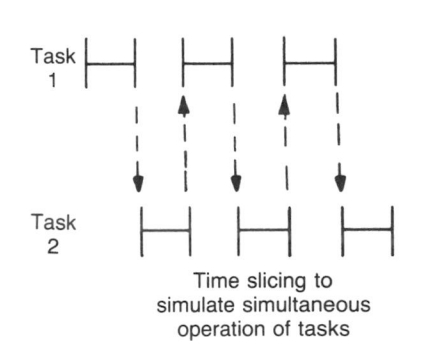

Time slicing to
simulate simultaneous
operation of tasks

However, one task might come to
a place where it needs to get some
information from another task or
to give some to another task. But
that other task may not be at the
right place so they can exchange
information without confusion.
So the first task waits until the

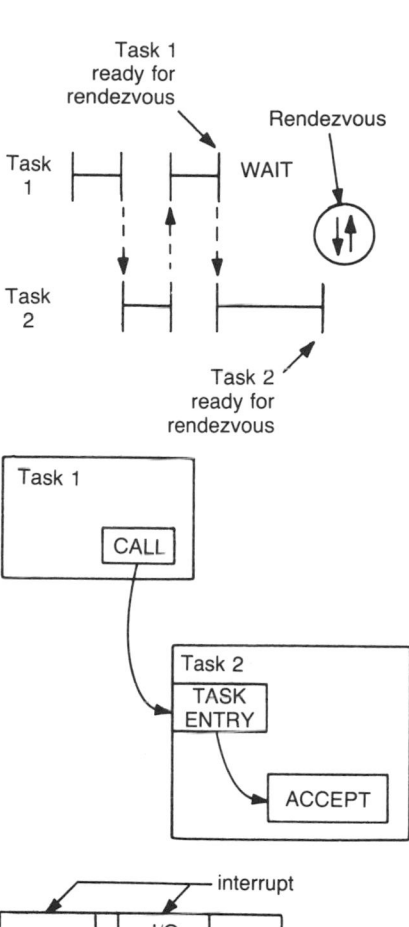

second task is ready, and then they get together, in what is called a "rendezvous." The task that got to the rendezvous first is suspended until the other one is ready, and during that time it will not receive its time slices.

When we want a task to ask for a rendezvous, we put into it a "call" to a "task entry" of the second task. In the second task there will be an "accept" statement, which denotes the correct place at which the second task can accommodate the rendezvous, by "accepting the call" to the task entry.

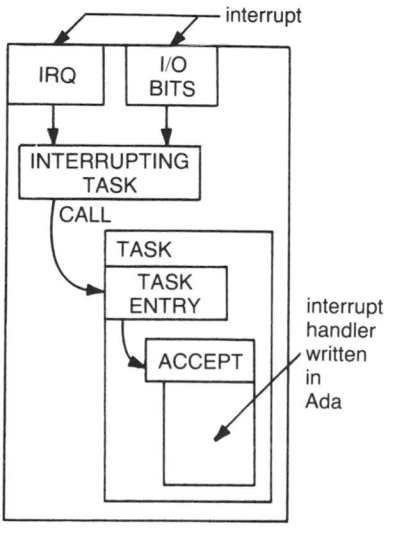

Interrupt handlers, when written in Ada, are generally set up so that they can accept a task entry call that is generated (we'll tell you how below) when the interrupt comes in. So the interrupt handler either must be a task by itself, or it must be inside a task. In our example, we've elected to put the handlers for both the interval timer interrupt (the clock ticks) and the comparator interrupt into our main task. One of

these task entries is called
"CLOCK_TICK" and the other
is called "COMPARATOR_SIG-
NAL." Each of these is mentioned
in an ACCEPT statement, which
is the correct place for the rendezvous with the interrupting task to
take place.

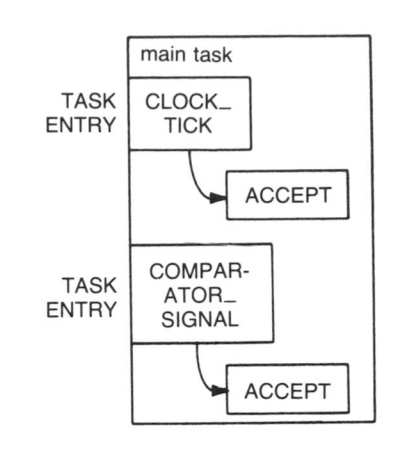

But just what is this interrupting
task? Is it something that we
wrote in Ada, or is it some other
code that the Ada compiler set
up? Actually, it's the "initial interrupt handler" that we described above. We don't have to
write it; it's generated by the Ada
compiler, though in some imple-
mentations it may call on service
routines that exist in ROM, to
help it do the job. The "initial interrupt handler" responds to the
interrupt (as soon as it happens),
determines which interrupt it is,
and then sets up a task entry call
to the correct Ada handler that
we have written. Then it gives
control back to the Ada.

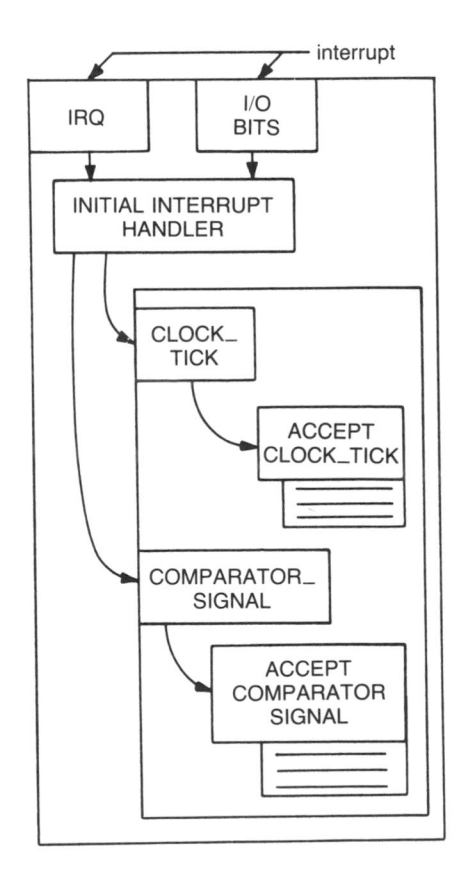

At that point, what happens is
again dependent on the implementation. One implementation
might recognize that the interrupt
was meant for a certain task, so
it will give control immediately
to that task. Another implementa-

tion might give control back to whatever task was taking its time slice when the interrupt came in, and let the time slicing go along until the right task to handle the interrupt comes along in its own time.

In either case, control eventually reaches the task that contains the ACCEPT statement. If that task is "sitting" at the ACCEPT statement, then it accepts the entry call immediately and begins to respond to the interrupt. But if it's not at the ACCEPT statement, then it continues executing the regular code of the task until it gets to the ACCEPT statement.

You might say that this would insert undesirable delays in the response to interrupts. And you would be right — Ada as it stands today doesn't do an outstanding job of responding quickly to interrupts. However, in our case, we've structured the task that contains the two interrupt handlers so that the delay will be very short and, in fact, small compared to the time between clock ticks — so there won't be a problem.

Ada compilers have been maturing, and interrupt handling in

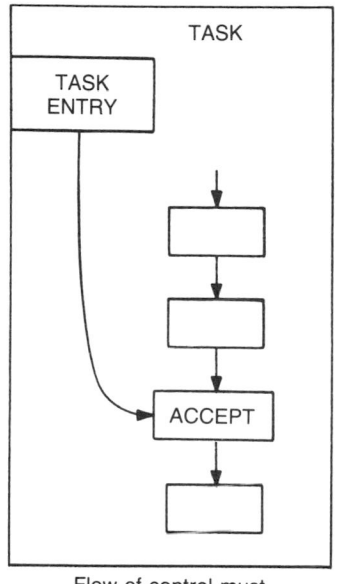

Flow of control must reach the correct ACCEPT statement before the task entry call will be accepted.

Ada is gradually getting better. Although we didn't need to do this in our example, some interrupt handlers might have to be divided into an "immediate" part and a "delayed" part. The purpose of that would be to take care of a short, critical part of the job right away, but delay any time-consuming part until other important parts of the Ada program were completed for the moment. The "immediate" part would be executed immediately after the "initial interrupt handler" we talked about above and before control returns to the Ada program. The "delayed" part would then be part of an ACCEPT statement in an Ada task, using the method we've used here, and it would be executed some time later. Because it would be in the ACCEPT statement, it wouldn't execute until the Ada program "wanted" it to.

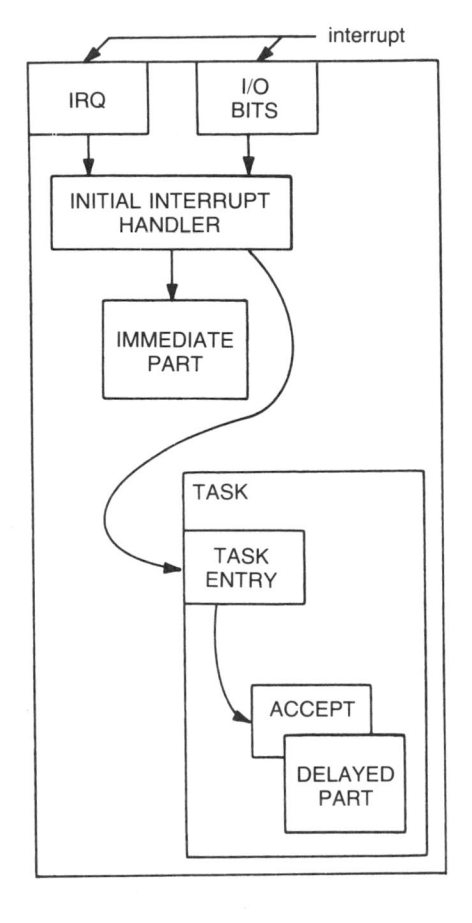

This "immediate" part could be written in Ada or in some other language, such as assembler. Either way, it could require special compiler features to make sure that it is put into memory correctly and will run when intended.

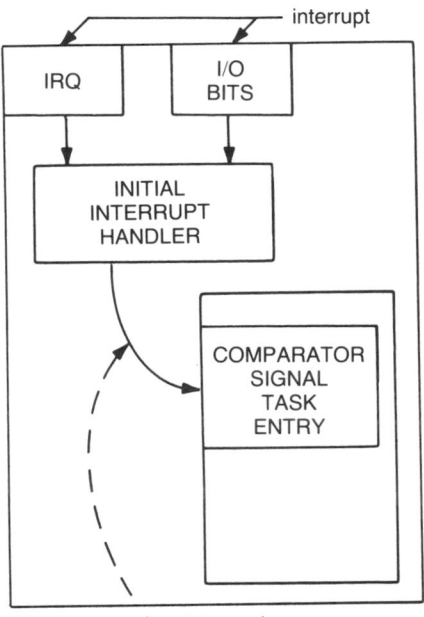

interrupt

IRQ

I/O BITS

INITIAL INTERRUPT HANDLER

COMPARATOR SIGNAL TASK ENTRY

address clause
makes this connection

Let's look at the interrupt we set up for the comparator. We defined "COMPARATOR_SIGNAL" as one of the task entries. To connect this name with the appropriate interrupt, we use an "address clause," such as:

for COMPARATOR_SIGNAL use at 16#01#;

where the hexadecimal address 01 is directly associated with the particular interrupt. This will tell the compiler what it needs to know in order to connect the task entry point "COMPARATOR_SIGNAL" with the correct interrupt.

With this connection, our Ada code can accept the call generated by the comparator interrupt. If the call is not present at that time, then our Ada code will take the alternate route within the "select" statement, accept the clock

203

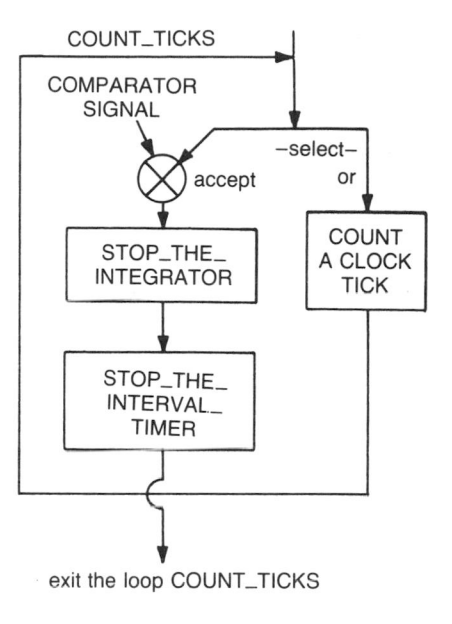

exit the loop COUNT_TICKS

tick interrupt if it's there, and go count another clock tick.

However, Ada requires that we identify this alternate route with a condition, and the condition is that we haven't gotten the comparator entry call yet. We set up the condition in this way:

Ada has an internal mechanism that is able to count the number of calls to a particular task entry and to provide that count for use in the program. To get at it, we use the special term 'count, appended to the name of the task entry: "COMPARATOR_SIGNAL'count." If there have been no calls to the task entry COMPARATOR_SIGNAL yet (that is, if the comparator signal has not been received yet), then COMPARATOR_SIGNAL'count will be zero, and that fact can be used. Ada has a number of built-in functions of this type.

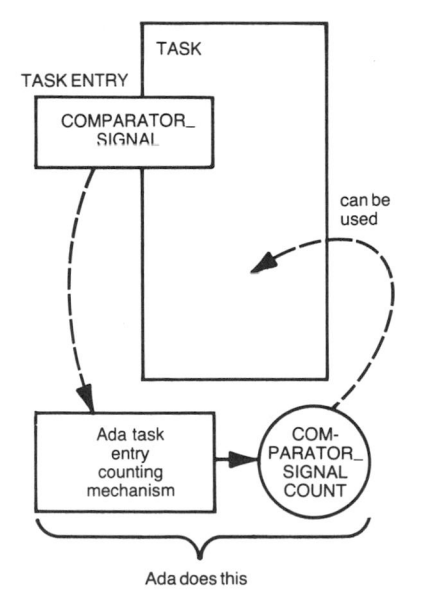

The rest of the loop COUNT_
TICKS is, then, as follows:

```
or when
COMPARATOR_SIGNAL 'count = 0
accept CLOCK_TICK;
CLOCK_TICK_COUNT:=

CLOCK_TICK_COUNT + 1;
if CLOCK_TICK_COUNT >
MAX_COUNT then
STOP_THE_INTEGRATOR;
STOP_THE_INTERVAL_TIMER;
PUT_THE_ERROR_DIGIT_CODES_
INTO_THE (READOUT);
SEND_TO_THE_DISPLAY_THE
(READOUT);
exit NORMAL_OPERATION;
enf if;
end select;
end loop COUNT_TICKS:
```

As you can see, this Ada code is
quite readable. Here, if the comparator signal is not present, we
accept the other task entry called CLOCK_TICK, which comes
from the interrupt that comes
from the interval timer's ticks.
This takes the place of the timing loop we employed in the assembly language version. When
the timer sends out its signal, the
variable   CLOCK_TICK—
COUNT is increased by 1, and
then a check is made to be sure
that this count hasn't gone over
its prescribed limit (if it has, error codes are sent to the display). Note that if CLOCK_
TICK_COUNT doesn't go over
that limit, control passes down

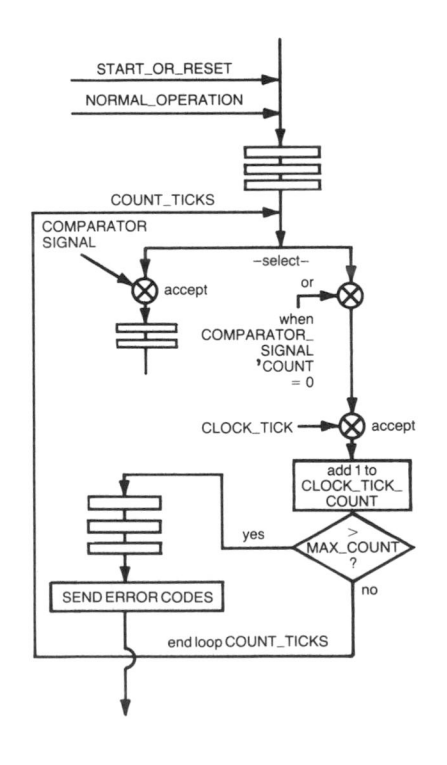

to the "end select" statement, and from there it automatically loops back to the beginning of the COUNT_TICKS loop. That causes the task to look again for the comparator signal to see if it has come in, before it again checks for the interval timer's CLOCK_TICK. If the program reaches the select statement before either signal arrives, then it will wait at the beginning of the statement until one does arrive.

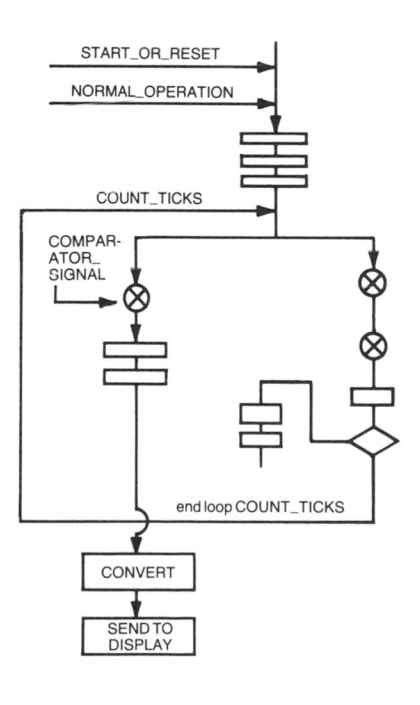

Suppose the comparator signal does come in. According to the code above, we then "exit COUNT _TICKS." That sends the control down past the "end loop COUNT _TICKS" statement. And there, we have placed this code:

CONVERT (CLOCK_TICK_COUNT, READOUT);

This is a call to the procedure CONVERT, providing as input to it the variable CLOCK_TICK_ COUNT, and getting back an answer called READOUT, which will contain the digit codes for the display. The program continues:

```
SEND_TO_THE_DISPLAY_THE
  (READOUT);
end loop NORMAL_OPERATION;
end loop START_OR_RESET;
end CONTROL_TASK;
```

The procedure "SEND_TO_THE_DISPLAY_THE" is the one which gets the digit codes contained in the variable "READOUT" to the displays. When control reaches the statement "end loop NORMAL_OPERATION," it will loop back to reset and start the integrator, start the interval timer, set the clock tick count to zero again, and begin the cycle over.

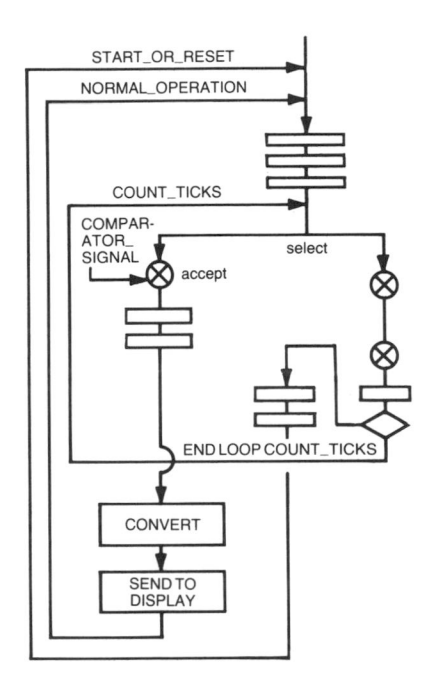

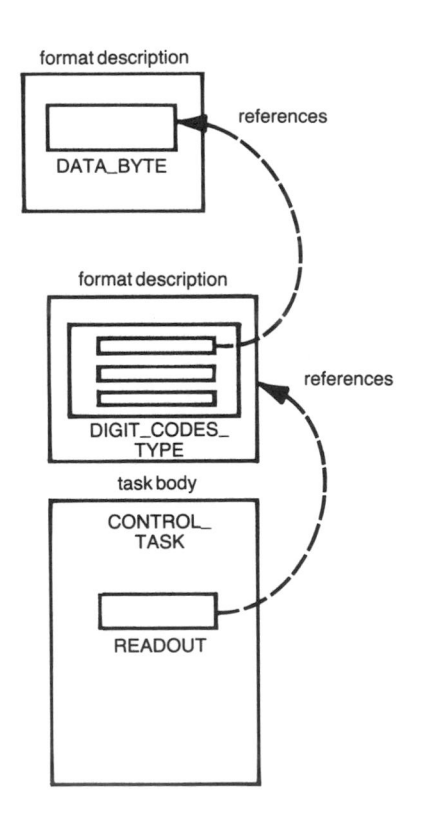

format description

DATA_BYTE

references

format description

DIGIT_CODES_
TYPE

references

task body

CONTROL_
TASK

READOUT

We have said that Ada requires various things to be defined in parts of the program separated from the executable code, so that they can be found easily if a change is needed. For instance, just ahead of all the code we have explained above, there is the following:

```
task body CONTROL_TASK is
READOUT: DIGIT_CODES_TYPE;
CLOCK_TICK_COUNT: integer range 0..300;
begin
```

and, much earlier in the program, we will find:

```
type DIGIT_CODES_TYPE is record
TENS_DIGIT_CODE: DATA_BYTE;
UNITS_DIGIT_CODE: DATA_BYTE;
TENTHS_DIGIT_CODE: DATA_BYTE;
end record:
```

and, even earlier in the program,

```
type DATA_BYTE is range 0..255;
for DATA_BYTE size use 8;
```

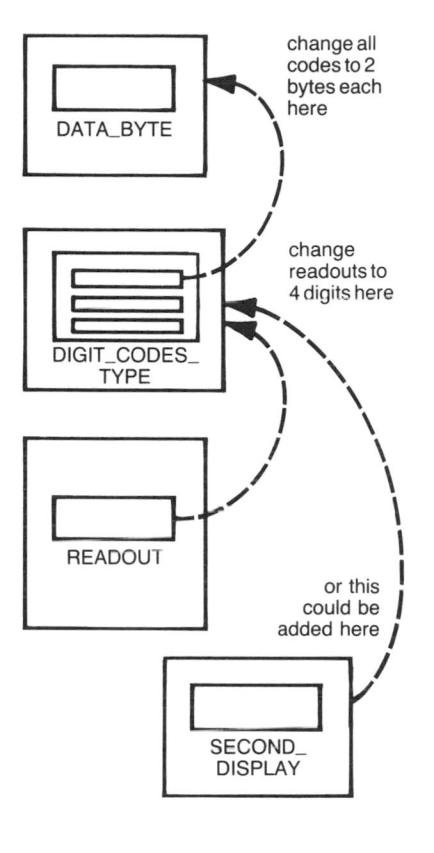

change all
codes to 2
bytes each
here

DATA_BYTE

change
readouts to
4 digits here

DIGIT_CODES_
TYPE

READOUT

or this
could be
added here

SECOND_
DISPLAY

This defines DATA_BYTE as an eight-bit byte (and forces the Ada compiler to implement it that way). Then it defines the three digit codes as being one byte each, and then gives them a collective name—first as a "type," DIGIT_CODES_TYPE"—and then (by using that name) defines an entity "READOUT" as a collection of the three named digit codes. Now, if there is ever a time when the basic digit code needs to be changed to, say, two bytes, there is one place where it can be done. If we ever need to add a fourth digit code (say, called "HUNDREDTHS_DIGIT-CODE"), there is one place to do that. And if we ever needed a second group of digit codes for any reason (say, to drive a second display with a different reading), we can define it using the type "DIGIT_CODES_TYPE" just the way we defined "READOUT" above. So Ada really does separate out the various areas where changes could occur.

At the end of all this, we'll show you the whole Ada program, in just the way the compiler expects to receive it. For now, though, let's look at another of the procedures, one called "RESET_ AND_START_THE_INTEGRA-TOR." That will illustrate some more features of Ada.

30 "PACKAGES" IN OUR EXAMPLE

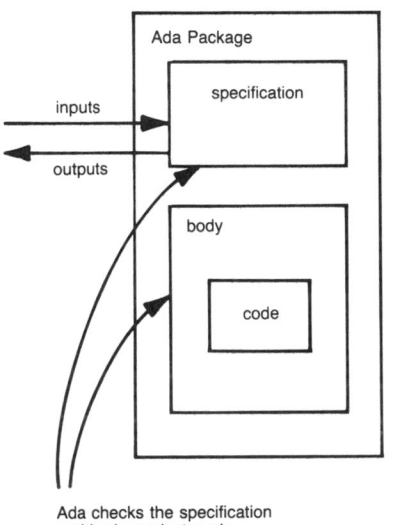

Ada Package

inputs

specification

outputs

body

code

Ada checks the specification and body against each other for consistency

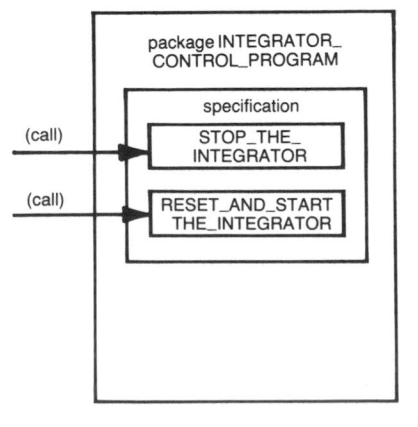

package INTEGRATOR_ CONTROL_PROGRAM

specification

(call)   STOP_THE_ INTEGRATOR

(call)   RESET_AND_START THE_INTEGRATOR

We said above that the "package" concept in Ada is based on the idea that it should be possible to isolate the definition of the inputs and outputs of a section of the program from the executable code, so that the programmers of other sections don't need to concern themselves with the first section's code (and are thereby prevented from causing the code harm). The definition of the inputs and outputs is done in the package "specification," and the executable code is put into what is called the package "body."

In our Ada program we have a package called INTEGRATOR_ CONTROL_PROGRAM:

package INTEGRATOR_CONTROL_PROGRAM is
procedure STOP_THE_INTEGRATOR;
procedure
RESET_AND_START_THE_INTEGRATOR;
end INTEGRATOR_CONTROL_PROGRAM;

The above is just the package "specification." It says that there are two procedures (routines) that other programs can call, and (as it happens) neither one of them has any data passed into it or out of it. All that the other programmers need to know are the names of the procedures, and that is all that is given.

The package "body," however, has the specific executable code for these two procedures, plus some code to cause initialization. Remember that in the assembly language version, the control of the integrator was done through one of the programmable interface adapters (PIA2). So the package body for the integrator control looks like this:

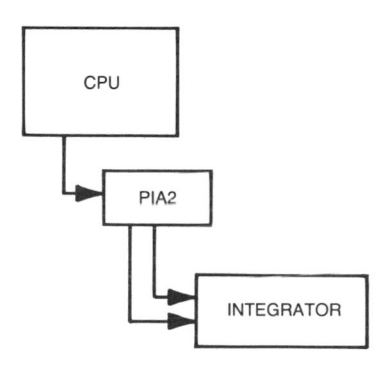

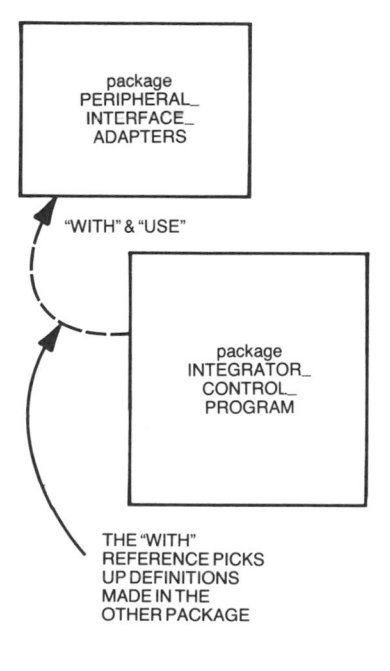

```
with PERIPHERAL_INTERFACE_ADAPTERS;
use PERIPHERAL_INTERFACE_ADAPTERS;

package body
INTEGRATOR_CONTROL_PROGRAM is

procedure STOP_THE_INTEGRATOR is
begin
 PIA2.SIDE_B_DATA:=16#01#;
end;

procedure
RESET_AND_START_THE_INTEGRATOR is
begin
 PIA2.SIDE_B_DATA:=16#00#;
end;

begin
 PIA2.SIDE_B_DATA:= SET_FOR_OUTPUT;
end INTEGRATOR_CONTROL_PROGRAM;
```

Here, the integrator control program is going to refer to some variables and constants defined in another package called PERIPHERAL_INTERFACE_ADAPTERS; so the purpose of the "with" and "use" clauses is to make those references possible. Without the "with" and "use," the programmer of this routine is deliberately prevented from making those references, and he is forced to do something special and highly visible (the "with" and "use") to get at that other package. That in turn will usually prevent him from doing unintended damage to the other package.

The two procedures above just set the variable PIA2.SIDE_B_DATA to hex 00 or hex 01, respectively. Before either of those procedures is called, however, we want the PIA side B register set to hex FF; that's the function of the "begin-end" block at the bottom. When we come to the package covering the PIA's, we will see that "SET _FOR_OUTPUT" has been defined as a byte containing hex FF. Keeping the definition of "PIA2.SIDE_B_DATA" in the package having to do with the PIAs is another way that we can keep things having to do with the

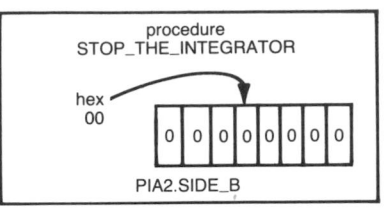

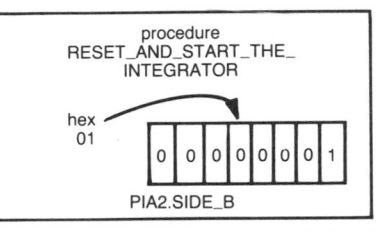

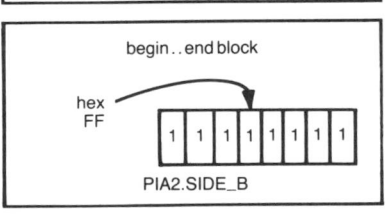

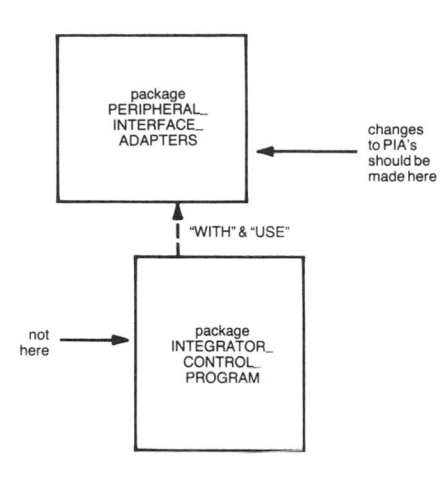

changes to PIA's should be made here

not here

PIAs together, so that if a change happens to them (say, a design change to the PIA chip) we can deal with them all in one place. The above routine having to do with the integrator won't be affected—and that's just what we wanted.

Here is the package having to do with the PIAs. First, the package specification begins as follows:

with system, COMMON_DECLARATIONS;
use COMMON_DECLARATIONS;

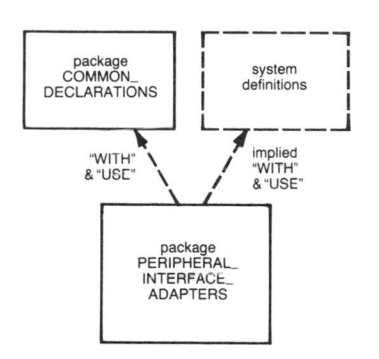

This allows us to refer to some definitions present in COMMON _DECLARATIONS (which we will have written) and some present in "system," a set of definitions that the Ada compiler will have made available to us. The latter is important because some of the actual machine addresses we will need to work with will be dependent on how the compiler has been written and for what machine it is translating code.

package PERIPHERAL_INTERFACE_ADAPTERS
is

SET_FOR_OUTPUT: constant DATA_BYTE :=
16#FF#;

SET_FOR_PERIPHERAL_REGISTERS : constant
DATA_BYTE:=16#04#;

SET_FOR_PERIPH_REG_AND_INT_CNTL :
constant DATA_BYTE:=16#07#;

```
ADDRESS_OF_PIA1 : constant system.address :=
16#0400#;

ADDRESS_OF_PIA2 : constant system.address :=
16#0800#;

BYTE:constant integer:=1;
```

In the above definition, the type "DATA_BYTE" has been defined in "COMMON_DECLARA-TIONS," and "system.address" is a type defined within the compiler.

What we are trying to do now is to set up the exact PIA register format so that it matches the way the PIA hardware is designed. The internal formats for PIA1 and PIA2 are the same, so we set them up as a "type" format description:

```
type PIA is
record
SIDE_A_DATA,
SIDE_A_CONTROL,
SIDE_B_DATA,
SIDE_B_CONTROL:
COMMON_DECLARATIONS.DATA_BYTE;
end record;
```

Here we define the data and control registers of the PIAs as four one-byte fields in memory. The Ada language, however, does not guarantee the exact positions of these bytes in memory; for example, each byte might need to be located at a word boundary. So we use a "representation clause"

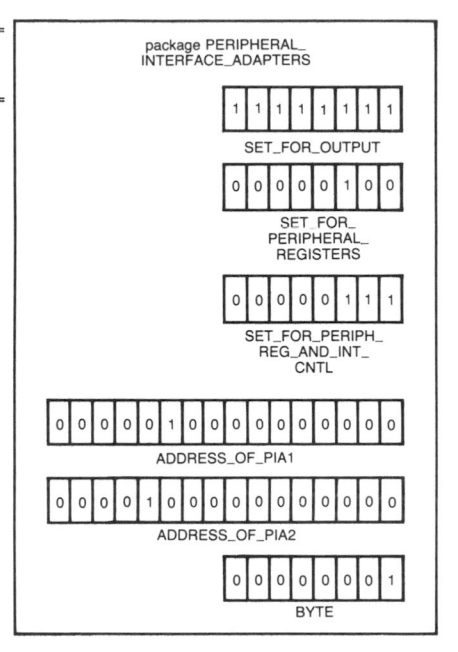

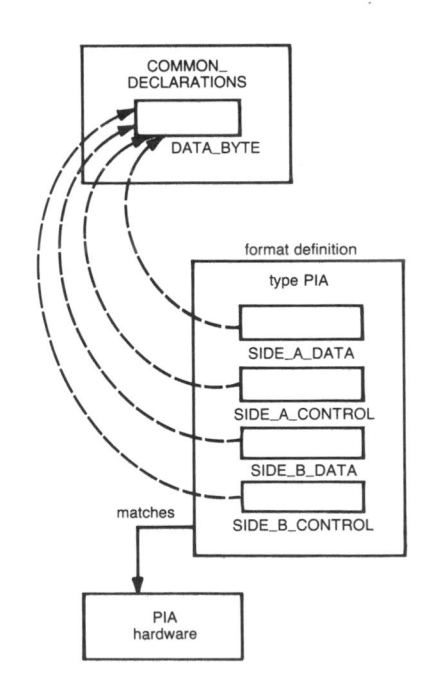

to force the locations of the bytes:

```
for PIA use

record at mod 4;
    SIDE_A_DATA        at 0*BYTE range 0..7;
    SIDE_A_CONTROL     at 1*BYTE range 0..7;
    SIDE_B_DATA        at 2*BYTE range 0..7;
    SIDE_B_CONTROL     at 3*BYTE range 0..7;
end record;
```

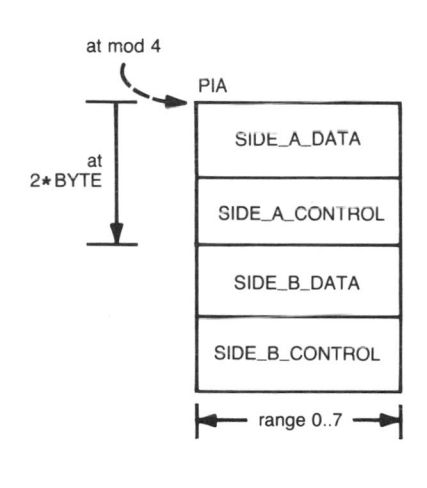

The "at mod 4" clause aligns the record to start at a memory address divisible by 4. The restrictions "at 0*BYTE," "at 1*BYTE," and so on give the precise offset of each field (such as "SIDE_A_CONTROL") within the record. The indication "range 0..7" assigns all eight bits of the byte to this field. Actually, the "range" part of a clause like this can be used to assign different bits of a single memory location to different fields of a record, so there's a great deal of flexibility in handling bits and bytes. Ada was set up this way so that programmers would not have to resort to the use of assembly language when doing work of that sort.

Now:

```
PIA1:PIA:=(others =>0);
for PIA1 use at ADDRESS_OF_PIA1;

PIA2:PIA:=(others =>0);
for PIA2 use at ADDRESS_OF_PIA2;

end PERIPHERAL_INTERFACE_ADAPTERS;
```

Here, we have defined the specific PIAs "PIA1" and "PIA2," defining them as type PIA (with all the attributes we described for the type PIA). We have tied each PIA to a definite starting address, which we defined earlier in the "PERIPHERAL_INTERFACE_ ADAPTERS" package. Now the locations of all the PIA registers are fully defined as definite locations in memory. Moreover, things have been split out so that if anything changes (the structure of the PIA registers, what memory addresses we want to wire them up for, how the bits are mapped in each register, what range of numbers each register can hold, and so on), each of these facts can be found independently and changed easily without damaging any other code.

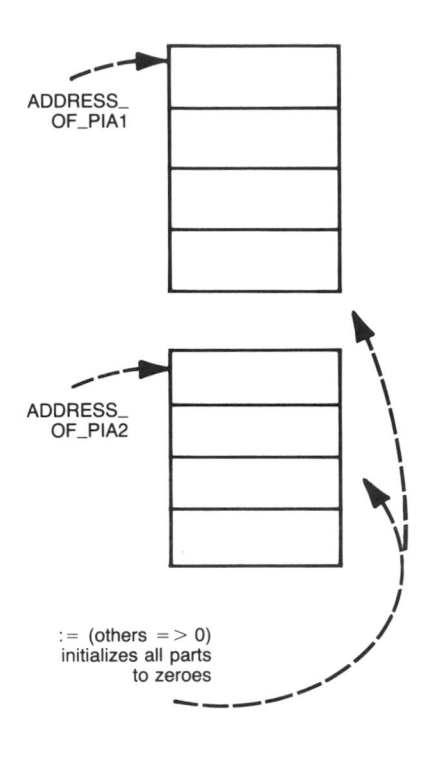

ADDRESS_
OF_PIA1

ADDRESS_
OF_PIA2

:= (others => 0)
initializes all parts
to zeroes

31

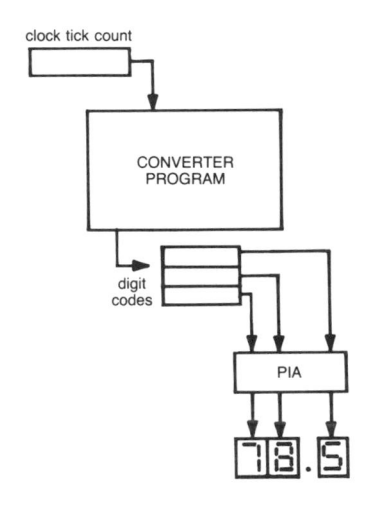

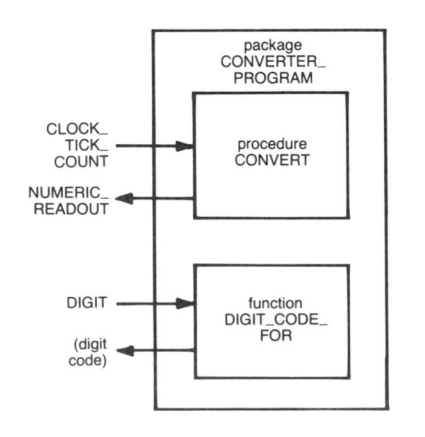

Now, let's look at the converter program. Remember that the input to the converter program is the clock tick count (a number from 0 to 199), and the output is a set of three digit codes, each of which is a byte that can be sent through a PIA to the seven-segment displays—with each bit of the byte driving one of the segments. So we say:

```
package CONVERTER_PROGRAM is

procedure CONVERT (CLOCK_TICK_COUNT:
  VALID_CLOCK_TICK_RANGE;
  NUMERIC_READOUT: out
  DIGIT_CODES_TYPE);

function DIGIT_CODE_FOR (DIGIT:integer)
  return DATA_BYTE;

end CONVERTER_PROGRAM;
```

This is just the specification for the package, and it is all that the programmer of the main control routine needs to know. How the conversion is actually done is the

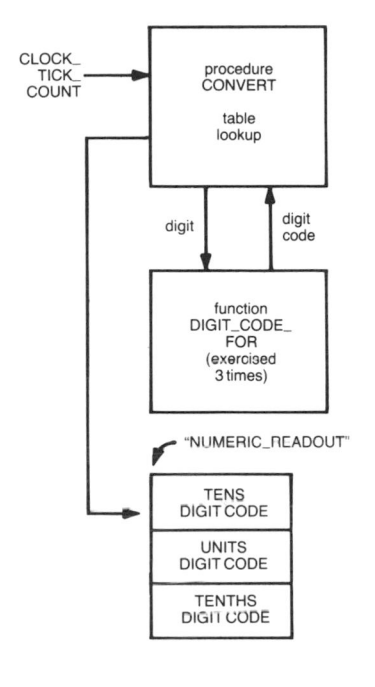

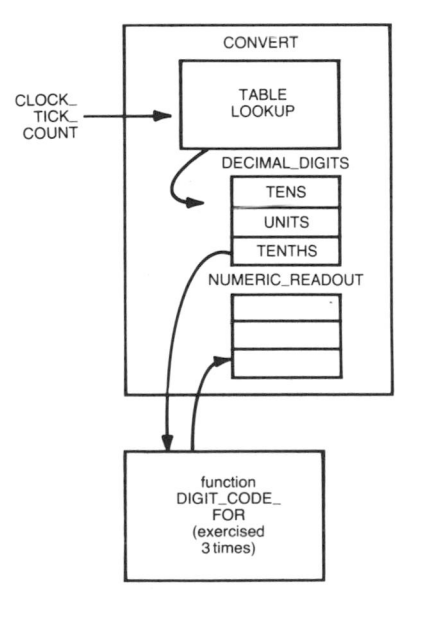

province of the programmer of CONVERTER_PROGRAM, who is the only one who needs to know about it.

We have structured CONVERTER _PROGRAM so that it does these two things separately: First, it looks up a set of three decimal digits in a table that represent the reading we want displayed. Second, it looks up (for each one of those digits) the digit code we need to send to the display. Doing it this way, if the readout hardware changes and the digit codes that we need also therefore change, they'll need to be changed in only one place. The procedure CONVERT does the first lookup, and then (from within itself) it calls the function DIGIT_CODE _FOR, which does the second lookup. In the body of package CONVERTER_PROGRAM, procedure CONVERT looks like this:

```
procedure CONVERT (CLOCK_TICK_COUNT:
    VALID_CLOCK_TICK_RANGE;
    NUMERIC_READOUT:out
    DIGIT_CODES_TYPE;

type DECIMAL_DIGITS_TYPE is
record
    TENS,
    UNITS,
    TENTHS: integer range 0..9;
end record;

DECIMAL_DIGITS: DECIMAL_DIGITS_TYPE;
```

This defines a type containing the three decimal digits collected into a record and then defines DECI-MAL_DIGITS as a variable name with which we can refer to the whole collection of three digits—yet refer to them individually when we want to. We would say, for instance, "DECIMAL_DIGITS .TENTHS," and that would pick out the third digit.

Next, we need to have the table itself:

```
TABLE_ENTRY_FOR_THE: array
   (VALID_CLOCK_TICK_RANGE)
   of
DECIMAL_DIGITS_TYPE:=
   (0 => (0,0,0),
    1 => (0,0,5),
    2 => (0,1,0),
    3 => (0,1,5),
```

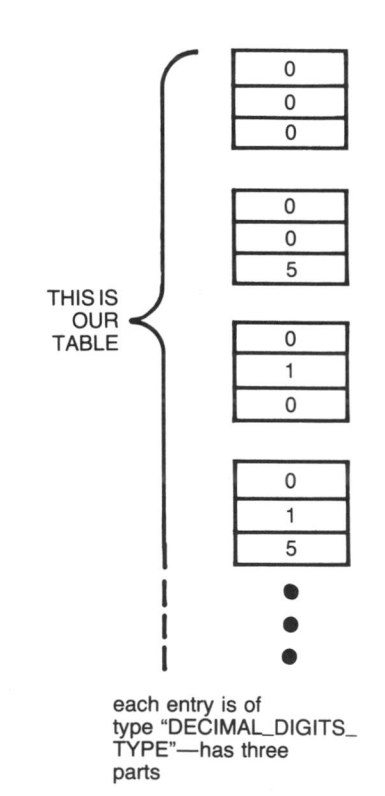

THIS IS
OUR
TABLE

each entry is of
type "DECIMAL_DIGITS_
TYPE"—has three
parts

(and so on, up to 199). Here we define an array (a table) where each entry of the table is of DE-CIMAL_DIGITS_TYPE, and then we specify the table itself. The notation 2 => (0,1,0) means that when the index to the table is 2, the entries in the table are the three separate decimal digits 0, 1, 0. This, of course, is only the beginning of the table (the entire table is shown in the program it-self, at the end of this discussion).

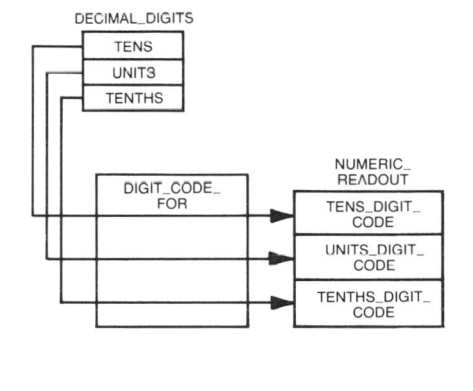

The actual work of the procedure CONVERT is done as follows:

```
begin
DECIMAL_DIGITS:=
TABLE_ENTRY_FOR_THE
(CLOCK_TICK_COUNT);
```

(This does the table lookup.)

```
NUMERIC_READOUT.TENS_DIGIT_CODE :=
DIGIT_CODE_FOR (DECIMAL_DIGITS.TENS);
```

(This converts the tens digit to a digit code.)

```
NUMERIC_READOUT.UNITS_DIGIT_CODE :=
DIGIT_CODE_FOR (DECIMAL_DIGITS.UNITS);

NUMERIC_READOUT.TENTHS_DIGIT_CODE
:=
DIGIT_CODE_FOR
(DECIMAL_DIGITS.TENTHS);

end CONVERT;
```

But, of course, the function DIGIT_CODE_FOR has to be defined:

```
function DIGIT_CODE_FOR (DIGIT:integer)
return DATA_BYTE is
begin
  case DIGIT is
    when 0 => return 2#01111110#;
    when 1 => return 2#00011000#;
    when 2 => return 2#01010111#;
    when 3 => return 2#01011101#;
    when 4 => return 2#00111001#;
    when 5 => return 2#01101101#;
    when 6 => return 2#01101111#;
    when 7 => return 2#01011000#;
    when 8 => return 2#01111111#;
    when 9 => return 2#01111101#;
    when others => return 2#01100111#;
  end case;

end DIGIT_CODE_FOR;
```

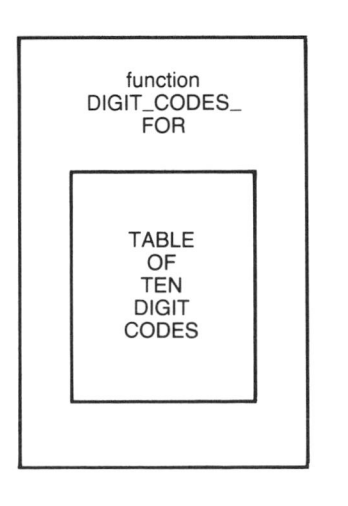

In the first line, we define the function and its input, an integer called DIGIT. A function, by definition, always returns a single variable; here it is of type DATA _BYTE, which we defined earlier this way:

```
type DATA_BYTE is range 0..255;
for DATA_BYTE 'size use 8;
```

—which defined it as a single eight-bit byte. The function is structured from a single "case" statement, which tells what to return in the eight-bit byte for each of the different cases of the input digit. The notation
       2#01100111#
means that this is a binary constant, made up of ones and zeroes. The last entry in the case statement says that when the input is not one of the digits 0 through 9, we return a digit code that brings an "E" (for error) up on the display.

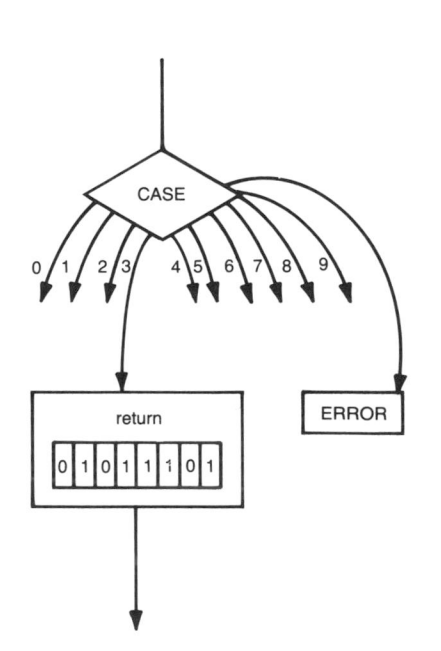

THE DISPLAY CONTROL PROGRAM

Remember that in the main control task we had a call to a procedure called "SEND_TO_THE_ DISPLAY_THE," which is used in this statement:

```
SEND_TO_THE_DISPLAY_THE(READOUT);
```

And remember also that "READOUT" was defined as being of type "DIGIT_CODES_TYPE," which in turn was defined as follows:

```
type DIGIT_CODES_TYPE is record
  TENS_DIGIT_CODE    :DATA_BYTE;
  UNITS_DIGIT_CODE   :DATA_BYTE;
  TENTHS_DIGIT_CODE  :DATA_BYTE;
end record;
```

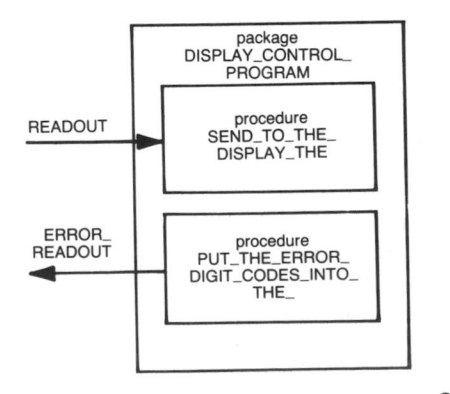

So SEND_TO_THE_DISPLAY _THE is going to send all three digit codes at once. But that procedure is only one procedure in a package we call "DISPLAY_ CONTROL_PROGRAM," which does some other things related to the display. The package is as follows:

with COMMON_DECLARATIONS;

use COMMON_DECLARATIONS;

package DISPLAY_CONTROL_PROGRAM is

procedure SEND_TO_THE_DISPLAY_THE
(READOUT: DIGIT_CODES_TYPE);

procedure
PUT_THE_ERROR_DIGIT_CODES_INTO THE
(ERROR_READOUT: out
DIGIT_CODES_TYPE);

end DISPLAY_CONTROL_PROGRAM;

The above is the package spec-ification; here is the body:

with PERIPHERAL_INTERFACE_ADAPTERS,
CONVERTER_PROGRAM;

use PERIPHERAL_INTERFACE_ADAPTERS,
CONVERTER_PROGRAM;

package body DISPLAY_CONTROL_PROGRAM
is
procedure SEND_TO_THE DISPLAY_THE
(READOUT:DIGIT_CODES_TYPE)
is
begin

PIA1.SIDE_A_DATA:=
READOUT.TENS_DIGIT_CODE;

PAI1.SIDE_B_DATA:=
READOUT.UNITS_DIGIT_CODE;

PAI2.SIDE_A_DATA:=
READOUT.TENTHS_DIGIT_CODE;

end SEND_TO_THE_DISPLAY_THE;

The above, of course, is what places the digit codes in the proper PIA registers. We con-tinue:

procedure
PUT_THE_ERROR_DIGIT_CODES_INTO_THE
(ERROR_READOUT:out DIGIT_CODES_TYPE)
is

```
ERROR:constant integer:= 10;
begin

ERROR_READOUT.TENS_DIGIT_CODE:=
DIGIT_CODE_FOR (ERROR);

ERROR_READOUT.UNITS_DIGIT_CODE:=
DIGIT_CODE_FOR (ERROR);

ERROR_READOUT.TENTHS_DIGIT_CODE):=
DIGIT_CODE_FOR (ERROR);

end;
```

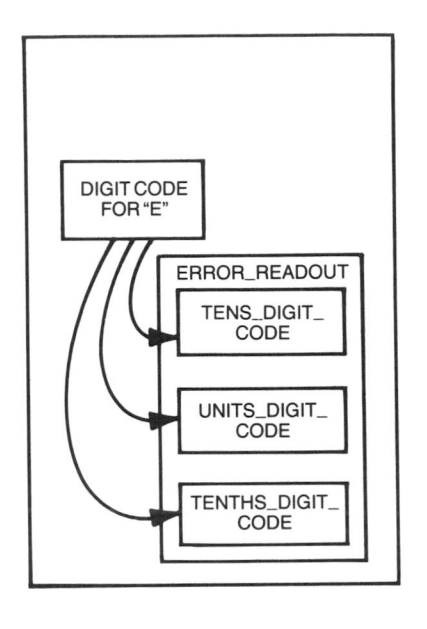

And the above code fills the display with "EEE." That takes care of defining the two procedures that are part of DISPLAY_CONTROL_PROGRAM, but we still need some code to initialize the PIAs when the program first starts up. The proper place for that code, in Ada, is at the end of the package:

```
PIA1.SIDE_A_DATA:=SET_FOR_OUTPUT;
PIA1.SIDE_B_DATA:=SET_FOR_OUTPUT;
PIA2.SIDE_A_DATA:=SET_FOR_OUTPUT;
```

(Back in the package PERIPHERAL_INTERFACE_ADAPTERS, we had defined SET_FOR_OUTPUT as constant DATA_BYTE:= 16#FF#, which made it equal to hex FF, or 1111 1111.)

```
PIA1.SIDE_A_CONTROL:=
SET_FOR_PERIPHERAL_REGISTERS;

PIA1.SIDE_B_CONTROL:=
SET_FOR_PERIPHERAL_REGISTERS;
```

(and back in PERIPHERAL_
INTERFACE_ADAPTERS, we
had defined SET_FOR_PERIPH-
ERAL_REGISTERS as constant
DATA_BYTE:= 16#04#, which
made it equal to hex 04, or 0000
0100.)

The remaining register, PIA2.
SIDE_A_CONTROL, is initial-
ized in the package COMPARA-
TOR_CONTROL_PROGRAM.

33 THE BENEFITS OF ADA— WITHOUT ADA

One thing that must be running through your mind by now is that we were right when we said that Ada syntax is complex and difficult, and it does take time to learn. Furthermore, there are other costs associated with Ada; Ada compilers are not available for every microprocessor (certainly not for the Motorola 6800, which is now a very old device), and so a microprocessor must be chosen for which a compiler exists. Those tend to be the more expensive microprocessors. Your application may not be able to afford them.

So — what if you like the idea of keeping things that need to be changed collected together in neat single places, keeping programs separated so that one programmer can't destroy another's code, mak-

ing the code readable and understandable, and so on? Is there a
way to get some of these benefits
without Ada?

It is really difficult to do those
things in assembly language. It's
an inherently unreadable language,
and new programmers must study
a program at length to understand
it. That can be true of languages
like Fortran as well, and partly
for the same reason: short names
for data and pieces of executable
code tend not to be self-explanatory. When they are limited to
four, six, or eight characters, they
just can't be as meaningful as,
say, a 20- or 30-character name
can be. There are other languages,
like "C" and Pascal, where that
restriction doesn't apply.

Any language in which you can
define types is better than one in
which you can't. None have quite
the "typing" power of Ada, but
don't let yourself be stuck with
just integer, floating point, and
characters. Types are one of the
ways you can coalesce things that
might change into single locations.
In languages with poor typing,
you can at least try to keep the

declarations or descriptions of your data items in one place, with comments that explain why those data items have been structured the way they are. That will help localize changes. A centralized list of definitions, at one point in the program, is useful in small projects, and when a large project is being developed by a number of programmers on a network, a common file can be used to store these definitions.

And just using the object idea, perhaps expressed as entity-relationship diagrams, or some other way of isolating pieces of software so that each corresponds to a single external entity or a single data structure where possible, will do a great deal to separate the pieces of your software. Changes will tend to occur at a single point, and programmers will not be as likely to destroy each other's code, since there will be a clearer division among the functions of different pieces of code.

But you can't get away from it — if you want the best of these benefits, you will have to use Ada. Ada wouldn't exist if that were not true, since Ada is difficult to learn.

Perhaps, in the future, Ada or languages even more advanced will become available for small, inexpensive microprocessors. When that happens, those design goals — and, really, business goals — of yours that have to do with error control and future change should be much easier to achieve.

34

THE ADA PROGRAM—
IN FINAL FORM

When the program was actually written, a number of the functions we had originally placed in certain software objects in the entity-relationship diagram were placed in other objects, because we felt that their presence in the new objects was more natural and therefore would help a new programmer to locate and change the code more easily. The acceptance of the comparator interrupt, for instance, was placed in the main control task, as were all the functions of the clock-tick program. Although we retained a separate interval timer control program, the clock tick was allowed to enter the main control task directly. In addition, the converter program and the display control program were permitted to communicate directly with one another. As you have seen, we created special software objects for

the common declarations and the declarations that had to do with the peripheral interface adapters.

With these changes, the portion of the entity-relationship diagram that is within (or represents) the computer now looks like the figure on page 237. Each block in this diagram is marked with the relevant page number in the program listing, which begins on page 238.

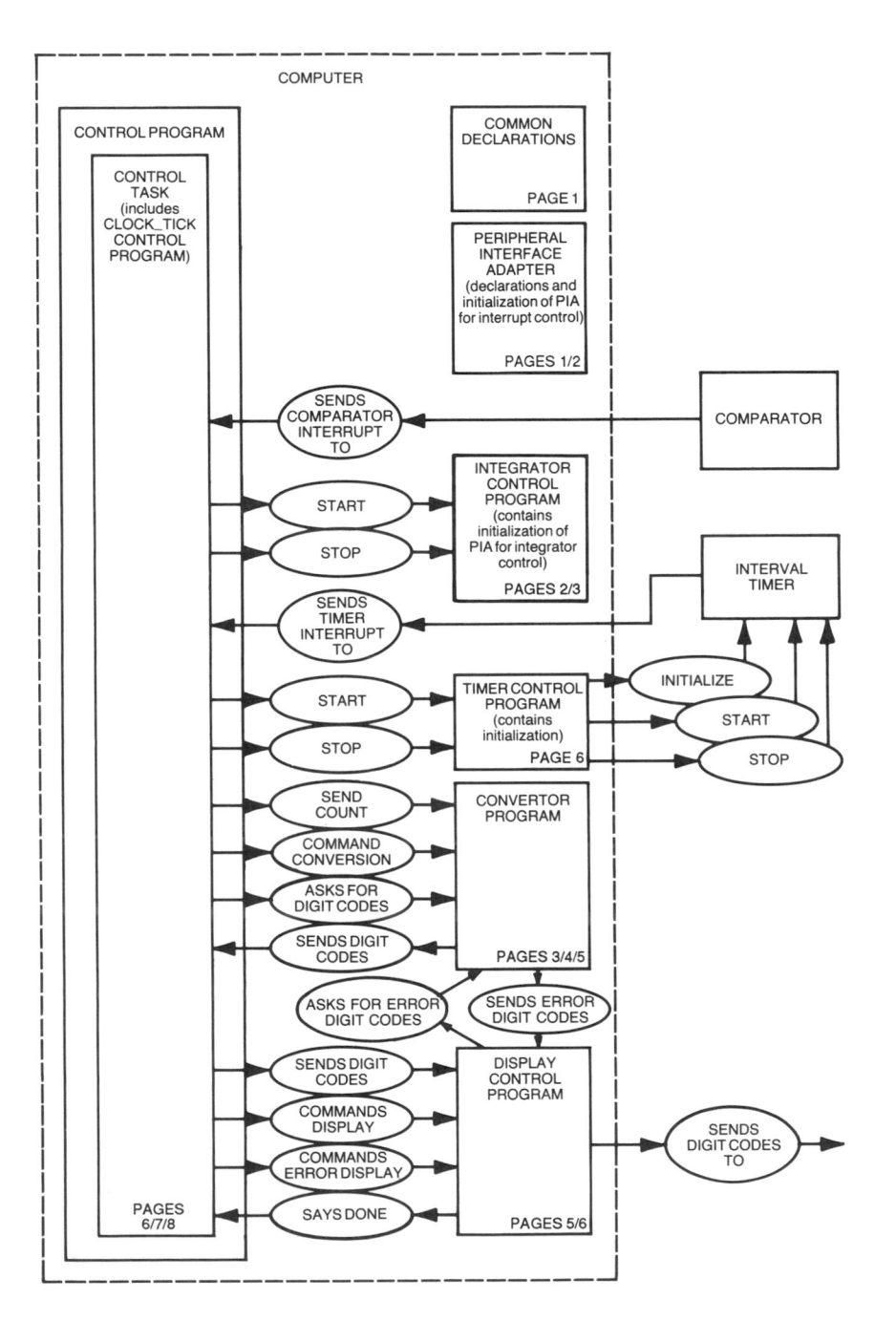

```
--=============================================================
-- PACKAGE SPECIFICATION: COMMON_DECLARATIONS
--
--    Types and variables which are used in several modules are declared here
--
--    MAX_COUNT is the number of clock ticks that represent a full tank
--
--    VALID_CLOCK_TICK_RANGE is the range of clock ticks between the values
--    representing an empty and a full tank
--
--    DATA_BYTE is an 8-bit byte that is required to be stored as a single byte in
--    memory
--
--    The type DIGIT_CODES_TYPE will contain the bit maps used to control the digital
--    display.  The 8-bit representation of each digit is necessary since it will be sent
--    directly to the hardware
--=============================================================

package COMMON_DECLARATIONS is

MAX_COUNT: constant integer := 200;

subtype VALID_CLOCK_TICK_RANGE is integer range 0..MAX_COUNT;

type DATA_BYTE is range 0..255;
for  DATA_BYTE'SIZE use 8;                        -- ensures that 8 bits of storage are used

type DIGIT_CODES_TYPE is record
    TENS_DIGIT_CODE:    DATA_BYTE;
    UNITS_DIGIT_CODE: DATA_BYTE;
    TENTHS_DIGIT_CODE: DATA_BYTE;
end record;

end COMMON_DECLARATIONS;

--=============================================================
-- PACKAGE SPECIFICATION: PERIPHERAL_INTERFACE_ADAPTERS
--
--    This package defines the type PIA (peripheral_interface_adapter).
--
--    It is represented as a record which needs to be mapped in a precise order to
--    sequential memory locations.
--
--    Two variables (PIA1 and PIA2) are declared, with corresponding address clauses to
--    place them correctly in memory.
--
--=============================================================

with system,
     COMMON_DECLARATIONS;
use COMMON_DECLARATIONS;

package PERIPHERAL_INTERFACE_ADAPTERS is

    SET_FOR_OUTPUT : constant DATA_BYTE := 16#FF#;
    SET_FOR_PERIPHERAL_REGISTERS : constant DATA_BYTE := 16#04#;
    SET_FOR_PERIPH_REG_AND_INT_CNTL : constant DATA_BYTE := 16#07#;
    ADDRESS_OF_PIA1 : constant system.address := 16#0400#; -- *
    ADDRESS_OF_PIA2 : constant system.address := 16#0800#; -- *
    --
    -- *  The type system.address is machine- and compiler-dependent
```

1

```
--
--      The values here correspond to the absolute machine addresses used in the
--      assembly code version of the example program and may not be appropriate for
--      the Ada implementation.
--
--      Ada compiler documentation will explain how to assign values to variables of type
--      system.address
--
BYTE : constant integer := 1;                              -- storage unit is one byte

type PIA is
      record
            SIDE_A_DATA,
            SIDE_A_CONTROL,
            SIDE_B_DATA,
            SIDE_B_CONTROL : COMMON_DECLARATIONS.DATA_BYTE;
      end record;

for PIA use
      record at mod 4;
            SIDE_A_DATA at 0*BYTE range 0..7;
            SIDE_A_CONTROL at 1*BYTE range 0..7;
            SIDE_B_DATA at 2*BYTE range 0..7;
            SIDE_B_CONTROL at 3*BYTE range 0..7;
      end record;

   PIA1 : PIA := (others => 0);                 -- all registers initialized to zero
   PIA2 : PIA := (others => 0);                 -- all registers initialized to zero

end PERIPHERAL_INTERFACE_ADAPTERS;

--==========================================================================
-- PACKAGE BODY: PERIPHERAL_INTERFACE_ADAPTERS
--==========================================================================

package body PERIPHERAL_INTERFACE_ADAPTERS is

begin
      --      initialization only:
      --      set appropriate control register for interrupt control and for data output to displays

      PIA2.SIDE_A_CONTROL := SET_FOR_PERIPH_REG_AND_INT_CNTL;

end PERIPHERAL_INTERFACE_ADAPTERS;

--==========================================================================
-- PACKAGE SPECIFICATION: INTEGRATOR_CONTROL_PROGRAM
--==========================================================================

package INTEGRATOR_CONTROL_PROGRAM is

      procedure STOP_THE_INTEGRATOR;

      procedure RESET_AND_START_THE_INTEGRATOR;

end INTEGRATOR_CONTROL_PROGRAM;

--==========================================================================
-- PACKAGE BODY: INTEGRATOR_CONTROL_PROGRAM
--
--      The package body will initialize the PIA2 so that SIDE B will control the integrator.
```

2

```
================================================================================--
                                                                                --
with  PERIPHERAL_INTERFACE_ADAPTERS;
use  PERIPHERAL_INTERFACE_ADAPTERS;

package body INTEGRATOR_CONTROL_PROGRAM is

    procedure STOP_THE_INTEGRATOR is
    begin
          PIA2.SIDE_B_DATA := 16#00#;
    end;

    procedure RESET_AND_START_THE_INTEGRATOR is
    begin
          PIA2.SIDE_B_DATA := 16#01#;
    end;

begin

    --    initialize PIA2, SIDE B for integrator control

    PIA2.SIDE_B_DATA := SET_FOR_OUTPUT;

end INTEGRATOR_CONTROL_PROGRAM;
--===========================================================================
-- PACKAGE SPECIFICATION: CONVERTER PROGRAM
--===========================================================================

with  COMMON_DECLARATIONS;
use  COMMON_DECLARATIONS;

package CONVERTER_PROGRAM is

    procedure CONVERT (CLOCK_TICK_COUNT : VALID_CLOCK_TICK_RANGE;
                       -- into
                            NUMERIC_READOUT : out DIGIT_CODES_TYPE);

    function DIGIT_CODE_FOR (DIGIT : integer) return DATA_BYTE;

end CONVERTER_PROGRAM;


--===========================================================================
-- PACKAGE BODY: CONVERTER PROGRAM
--
--    Note: The function DIGIT_CODE_FOR produces the bit-map for a single digit in the
--    display using the figure on page 33
--
--===========================================================================

package body CONVERTER_PROGRAM is

    function DIGIT_CODE_FOR (DIGIT : integer) return DATA_BYTE is
    begin
        case DIGIT is
            when 0 => return 2#01111110#;              -- see the figure on page 33
            when 1 => return 2#00011000#;
            when 2 => return 2#01010111#;
            when 3 => return 2#01011101#;
            when 4 => return 2#00111001#;
```

3

```
                    when 5 => return 2#01101101#;
                    when 6 => return 2#01101111#;
                    when 7 => return 2#01011000#;
                    when 8 => return 2#01111111#;
                    when 9 => return 2#01111101#;
                    when others => return 2#01100111#;              -- error character 'E'
                    --   in particular the ERROR value of 10, used by the procedure
                    --   PUT_THE_ERROR_DIGIT_CODES_INTO_THE, produces 'E'
              end case;
        end DIGIT_CODE_FOR;


        procedure CONVERT (CLOCK_TICK_COUNT : VALID_CLOCK_TICK_RANGE;
                           NUMERIC_READOUT  : out DIGIT_CODES_TYPE) is


        type DECIMAL_DIGITS_TYPE is
              record
                    TENS,
                    UNITS,
                    TENTHS : integer range 0..9;
              end record;


        DECIMAL_DIGITS : DECIMAL_DIGITS_TYPE;


        TABLE_ENTRY_FOR_THE : array (VALID_CLOCK_TICK_RANGE)
                                of DECIMAL_DIGITS_TYPE :=


        --   currently this table is linear, and it should be adjusted to the particular tank in the
        --   application


        (   0 => (0,0,0),   1 => (0,0,5),   2 => (0,1,0),   3 => (0,1,5),   4 => (0,2,0),
            5 => (0,2,5),   6 => (0,3,0),   7 => (0,3,5),   8 => (0,4,0),   9 => (0,4,5),
           10 => (0,5,0),  11 => (0,5,5),  12 => (0,6,0),  13 => (0,6,5),  14 => (0,7,0),
           15 => (0,7,5),  16 => (0,8,0),  17 => (0,8,5),  18 => (0,9,0),  19 => (0,9,5),
           20 => (1,0,0),  21 => (1,0,5),  22 => (1,1,0),  23 => (1,1,5),  24 => (1,2,0),
           25 => (1,2,5),  26 => (1,3,0),  27 => (1,3,5),  28 => (1,4,0),  29 => (1,4,5),
           30 => (1,5,0),  31 => (1,5,5),  32 => (1,6,0),  33 => (1,6,5),  34 => (1,7,0),
           35 => (1,7,5),  36 => (1,8,0),  37 => (1,8,5),  38 => (1,9,0),  39 => (1,9,5),
           40 => (2,0,0),  41 => (2,0,5),  42 => (2,1,0),  43 => (2,1,5),  44 => (2,2,0),
           45 => (2,2,5),  46 => (2,3,0),  47 => (2,3,5),  48 => (2,4,0),  49 => (2,4,5),
           50 => (2,5,0),  51 => (2,5,5),  52 => (2,6,0),  53 => (2,6,5),  54 => (2,7,0),
           55 => (2,7,5),  56 => (2,8,0),  57 => (2,8,5),  58 => (2,9,0),  59 => (2,9,5),
           60 => (3,0,0),  61 => (3,0,5),  62 => (3,1,0),  63 => (3,1,5),  64 => (3,2,0),
           65 => (3,2,5),  66 => (3,3,0),  67 => (3,3,5),  68 => (3,4,0),  69 => (3,4,5),
           70 => (3,5,0),  71 => (3,5,5),  72 => (3,6,0),  73 => (3,6,5),  74 => (3,7,0),
           75 => (3,7,5),  76 => (3,8,0),  77 => (3,8,5),  78 => (3,9,0),  79 => (3,9,5),
           80 => (4,0,0),  81 => (4,0,5),  82 => (4,1,0),  83 => (4,1,5),  84 => (4,2,0),
           85 => (4,2,5),  86 => (4,3,0),  87 => (4,3,5),  88 => (4,4,0),  89 => (4,4,5),
           90 => (4,5,0),  91 => (4,5,5),  92 => (4,6,0),  93 => (4,6,5),  94 => (4,7,0),
           95 => (4,7,5),  96 => (4,8,0),  97 => (4,8,5),  98 => (4,9,0),  99 => (4,9,5),
          100 => (5,0,0),101 => (5,0,5),102 => (5,1,0),103 => (5,1,5),104 => (5,2,0),
          105 => (5,2,5),106 => (5,3,0),107 => (5,3,5),108 => (5,4,0),109 => (5,4,5),
          110 => (5,5,0),111 => (5,5,5),112 => (5,6,0),113 => (5,6,5),114 => (5,7,0),
          115 => (5,7,5),116 => (5,8,0),117 => (5,8,5),118 => (5,9,0),119 => (5,9,5),
          120 => (6,0,0),121 => (6,0,5),122 => (6,1,0),123 => (6,1,5),124 => (6,2,0),
          125 => (6,2,5),126 => (6,3,0),127 => (6,3,5),128 => (6,4,0),129 => (6,4,5),
          130 => (6,5,0),131 => (6,5,5),132 => (6,6,0),133 => (6,6,5),134 => (6,7,0),
          135 => (6,7,5),136 => (6,8,0),137 => (6,8,5),138 => (6,9,0),139 => (6,9,5),
          140 => (7,0,0),141 => (7,0,5),142 => (7,1,0),143 => (7,1,5),144 => (7,2,0),
          145 => (7,2,5),146 => (7,3,0),147 => (7,3,5),148 => (7,4,0),149 => (7,4,5),
          150 => (7,5,0),151 => (7,5,5),152 => (7,6,0),153 => (7,6,5),154 => (7,7,0),
          155 => (7,7,5),156 => (7,8,0),157 => (7,8,5),158 => (7,9,0),159 => (7,9,5),
```

4

```
       160 => (8,0,0),161 => (8,0,5),162 => (8,1,0),163 => (8,1,5),164 => (8,2,0),
       165 => (8,2,5),166 => (8,3,0),167 => (8,3,5),168 => (8,4,0),169 => (8,4,5),
       170 => (8,5,0),171 => (8,5,5),172 => (8,6,0),173 => (8,6,5),174 => (8,7,0),
       175 => (8,7,5),176 => (8,8,0),177 => (8,8,5),178 => (8,9,0),179 => (8,9,5),
       180 => (9,0,0),181 => (9,0,5),182 => (9,1,0),183 => (9,1,5),184 => (9,2,0),
       185 => (9,2,5),186 => (9,3,0),187 => (9,3,5),188 => (9,4,0),189 => (9,4,5),
       190 => (9,5,0),191 => (9,5,5),192 => (9,6,0),193 => (9,6,5),194 => (9,7,0),
       195 => (9,7,5),196 => (9,8,0),197 => (9,8,5),198 => (9,9,0),199 => (9,9,5),
       200 => (9,9,9) );
   begin
       DECIMAL_DIGITS := TABLE_ENTRY_FOR_THE (CLOCK_TICK_COUNT);
       NUMERIC_READOUT.TENS_DIGIT_CODE
                  := DIGIT_CODE_FOR (DECIMAL_DIGITS.TENS);
       NUMERIC_READOUT.UNITS_DIGIT_CODE
                  := DIGIT_CODE_FOR (DECIMAL_DIGITS.UNITS);
       NUMERIC_READOUT.TENTHS_DIGIT_CODE
                  := DIGIT_CODE_FOR (DECIMAL_DIGITS.TENTHS);
   end CONVERT;

 end CONVERTER_PROGRAM;


-- PACKAGE SPECIFICATION: DISPLAY_CONTROL_PROGRAM


with COMMON_DECLARATIONS;
use  COMMON_DECLARATIONS;

package DISPLAY_CONTROL_PROGRAM is

   procedure SEND_TO_THE_DISPLAY_THE (READOUT : DIGIT_CODES_TYPE);

   procedure PUT_THE_ERROR_DIGIT_CODES_INTO_THE
                              (ERROR_READOUT : out DIGIT_CODES_TYPE);

end DISPLAY_CONTROL_PROGRAM;


-- PACKAGE BODY: DISPLAY_CONTROL_PROGRAM


with PERIPHERAL_INTERFACE_ADAPTERS,
     CONVERTER_PROGRAM;
use  PERIPHERAL_INTERFACE_ADAPTERS,
     CONVERTER_PROGRAM;

package body DISPLAY_CONTROL_PROGRAM is

   procedure SEND_TO_THE_DISPLAY_THE (READOUT : DIGIT_CODES_TYPE) is
   begin
       PIA1.SIDE_A_DATA := READOUT.TENS_DIGIT_CODE;
       PIA1.SIDE_B_DATA := READOUT.UNITS_DIGIT_CODE;
       PIA2.SIDE_A_DATA := READOUT.TENTHS_DIGIT_CODE;
   end SEND_TO_THE_DISPLAY_THE;

   procedure PUT_THE_ERROR_DIGIT_CODES_INTO_THE
                              (ERROR_READOUT : out DIGIT_CODES_TYPE) is
   ERROR : constant integer := 10;

   begin
```

```
            ERROR_READOUT.TENS_DIGIT_CODE := DIGIT_CODE_FOR (ERROR);
            ERROR_READOUT.UNITS_DIGIT_CODE := DIGIT_CODE_FOR (ERROR);
            ERROR_READOUT.TENTHS_DIGIT_CODE := DIGIT_CODE_FOR (ERROR);
      end;

      begin
            --    continue initialization of the PIAs, set data direction for output

                  PIA1.SIDE_A_DATA := SET_FOR_OUTPUT;
                  PIA1.SIDE_B_DATA := SET_FOR_OUTPUT;
                  PIA2.SIDE_A_DATA := SET_FOR_OUTPUT;

            -- set control so that data will go out to the displays

                  PIA1.SIDE_A_CONTROL := SET_FOR_PERIPHERAL_REGISTERS;
                  PIA1.SIDE_B_CONTROL := SET_FOR_PERIPHERAL_REGISTERS;

            -- PIA2.SIDE_A_CONTROL will be initialized in the package
            -- COMPARATOR_CONTROL_PROGRAM

      end DISPLAY_CONTROL_PROGRAM;
--=========================================================================
-- PACKAGE SPECIFICATION: TIMER_CONTROL_PROGRAM
--
--    An interval timer can be incorporated into the application which will send interrupts to
--    the CPU at a programmable rate.
--
--    This package is only used to initialize the timer.
--
--=========================================================================

package TIMER_CONTROL_PROGRAM is

      procedure START_THE_INTERVAL_TIMER;

      procedure STOP_THE_INTERVAL_TIMER;

end TIMER_CONTROL_PROGRAM;

--=========================================================================
-- PACKAGE BODY: TIMER_CONTROL_PROGRAM
--
--    Omitted
--
--    Would be similar to integrator_control_program.
--
--=========================================================================
-- MAIN PROCEDURE: CONTROL_PROGRAM
--
--    CONTROL_PROGRAM executes no instructions but instead spawns a dependent task
--    called the CONTROL_TASK which runs -- everything.
--
--    A task must be used since Ada will only interface to hardware interrupts through "task
--    entries".
--
--    The timer and comparator interrupts are connected to the two entries of the
--    CONTROL_TASK
--
--=========================================================================
```

6

```
with system,
    --   the system package must be visible if the hardware interrupt scheme of the Ada
    --   implementation utilizes system.address
    COMMON_DECLARATIONS,
    CONVERTER_PROGRAM,
    INTEGRATOR_CONTROL_PROGRAM,
    DISPLAY_CONTROL_PROGRAM,
    TIMER_CONTROL_PROGRAM,
    PERIPHERAL_INTERFACE_ADAPTERS;

use system,
    COMMON_DECLARATIONS,
    CONVERTER_PROGRAM,
    INTEGRATOR_CONTROL_PROGRAM,
    DISPLAY_CONTROL_PROGRAM,
    TIMER_CONTROL_PROGRAM,
    PERIPHERAL_INTERFACE_ADAPTERS;

procedure CONTROL_PROGRAM is

    task CONTROL_TASK is
        entry COMPARATOR_SIGNAL;
        for COMPARATOR_SIGNAL use at 16#01#; -- *
        entry CLOCK_TICK;
        for CLOCK_TICK use at 16#02#;        -- *
        -- * The interrupt values 01 and 02 are only illustrative.  The actual values
        --   possible will be dependent on the hardware configuration and the compiler
        --   used.  The Ada compiler documentation will explain how to use interrupt
        --   entries
    end CONTROL_TASK;

task body CONTROL_TASK is

    READOUT : DIGIT_CODES_TYPE;
    CLOCK_TICK_COUNT : integer range 0..300;

begin
    START_OR_RESET: loop
        NORMAL_OPERATION: loop

            CLOCK_TICK_COUNT := 0;
            RESET_AND_START_THE_INTEGRATOR;
            START_THE_INTERVAL_TIMER;

            COUNT_TICKS: loop
                select
                    --   if an interrupt has not arrived from the Comparator or
                    --   Timer the task will wait here.  If either or both interrupts
                    --   have arrived, an "accept" will be executed
                    accept COMPARATOR_SIGNAL;
                    STOP_THE_INTEGRATOR;
                    STOP_THE_INTERVAL_TIMER;
                    exit COUNT_TICKS;

                or when COMPARATOR_SIGNAL'count = 0
                    --   the attribute "count" is the number of calls to the named
                    --   task entry.  Here it is the number of outstanding interrupts
                    --   from the Comparator.  The value will be 0 or 1 in this
                    --   application.  When the Comparator has not signalled
                    --   we will accept a tick from the clock
```

7

```
                => accept CLOCK_TICK;
                       CLOCK_TICK_COUNT := CLOCK_TICK_COUNT + 1;

                   if CLOCK_TICK_COUNT > MAX_COUNT then
                       STOP_THE_INTEGRATOR;
                       STOP_THE_INTERVAL_TIMER;
                       PUT_THE_ERROR_DIGIT_CODES_INTO_THE
                                                    (READOUT);
                       SEND_TO_THE_DISPLAY_THE (READOUT);
                       exit NORMAL_OPERATION;
                   end if;

           end select;
        end loop COUNT_TICKS;

        -- At this point, the COUNT_TICKS loop must have
        -- terminated because of a Comparator interrupt

        CONVERT (CLOCK_TICK_COUNT,
                       -- into the
                               READOUT);
        SEND_TO_THE_DISPLAY_THE (READOUT);
      end loop NORMAL_OPERATION;
    end loop START_OR_RESET;
  end CONTROL_TASK;

begin -- procedure CONTROL
   --    CONTROL is a driver procedure which only exists to run the CONTROL_TASK.
   --    Therefore this procedure body is null
   null;
end CONTROL_PROGRAM;
```

8

INDEX

ABOUT THE AUTHORS

WILLIAM S. BENNETT is a Senior Member of the Technical Staff of the Link Flight Simulation Division of CAE-Link Corporation, Binghamton, New York. The author of some 25 articles on logic design and microcomputers, he received the B.S.E.E. (1952) degree from Carnegie-Mellon University, Pittsburgh, Pennsylvania.

CARL F. EVERT, JR., is Professor Emeritus in the Department of Electrical and Computer Engineering at the University of Cincinnati, Ohio, where he directed a microprocessor application laboratory and taught courses on computer-related topics. He is currently a consultant in private practice in computer technology, based in Cincinnati, Ohio. The author of many articles in electrical engineering and computer technology, Dr. Evert received the B.S. (1949) and M.S. (1950) degrees from Purdue University, Lafayette, Indiana, and Ph.D. (1959) degree from the University of Wisconsin—Madison.

LESLIE C. LANDER is Assistant Professor in the Department of Computer Science, The Thomas J. Watson School of Engineering, Applied Science, and Technology, State University of New York at Binghamton. The author or coauthor of several journal articles, reports, and a book, he is a member of the Association for Computing Machinery and an associate member of the Institute of Electrical and Electronics Engineers (Computer Society). Dr. Lander's research includes methodologies for the specification, design, and performance prediction of Ada-based real-time systems. He received the B.A. (1967) degree from Trinity College, University of Cambridge, England, and M.S. (1968) and Ph.D. (1973) degrees from the University of Liverpool, England.

Other volumes in preparation

WHAT EVERY ENGINEER SHOULD KNOW
A Series

Editor

William H. Middendorf

Department of Electrical and Computer Engineering
University of Cincinnati
Cincinnati, Ohio

WHAT EVERY ENGINEER
SHOULD KNOW ABOUT

MICROCOMPUTERS